Other titles in the *Inside Science* series:

Biotech Research
Climate Change Research
Gene Therapy Research
Infectious Disease Research
Mental Illness Research
Natural Disaster Research
Renewable Energy Research
Space Research
Sports Medicine Research
Stem Cell Research
Vaccine Research

Inside SCIENCE

Weapons and Defense Research

David Robson

San Diego, CA

© 2013 ReferencePoint Press, Inc.
Printed in the United States

For more information, contact:
ReferencePoint Press, Inc.
PO Box 27779
San Diego, CA 92198
www.ReferencePointPress.com

ALL RIGHTS RESERVED.
No part of this work covered by the copyright hereon may be reproduced or used in any form or by any means—graphic, electronic, or mechanical, including photocopying, recording, taping, web distribution, or information storage retrieval systems—without the written permission of the publisher.

LIBRARY OF CONGRESS CATALOGING-IN-PUBLICATION DATA

Robson, David, 1966-
  Weapons and defense research / by David Robson.
      p. cm. -- (Inside science)
  Includes bibliographical references and index.
  ISBN 978-1-60152-466-9 (hbk.) -- ISBN 1-60152-466-8 (hbk.) 1. Military weapons. 2. Military research. I. Title.
  UF500.R63 2013
  623.4072--dc23
                                                                          2012020754

Contents

Foreword	6
Important Events in Weapons and Defense Research	8
Introduction Confronting Threats in the Twenty-First Century	10
Chapter One What Is Weapons and Defense Research?	13
Chapter Two Drones	27
Chapter Three Cyberwarfare	40
Chapter Four Protecting Troops	52
Chapter Five The Future of Weapons and Defense Research	65
Source Notes	77
Facts About Weapons and Defense	82
Related Organizations	84
For Further Research	87
Index	89
Picture Credits	95
About the Author	96

Foreword

In 2008, when the Yale Project on Climate Change and the George Mason University Center for Climate Change Communication asked Americans, "Do you think that global warming is happening?" 71 percent of those polled—a significant majority—answered "yes." When the poll was repeated in 2010, only 57 percent of respondents said they believed that global warming was happening. Other recent polls have reported a similar shift in public opinion about climate change.

Although respected scientists and scientific organizations worldwide warn that a buildup of greenhouse gases, mainly caused by human activities, is bringing about potentially dangerous and long-term changes in Earth's climate, it appears that doubt is growing among the general public. What happened to bring about this change in attitude over such a short period of time? Climate change skeptics claim that scientists have greatly overstated the degree and the dangers of global warming. Others argue that powerful special interests are minimizing the problem for political gain. Unlike experiments conducted under strictly controlled conditions in a lab or petri dish, scientific theories, facts, and findings on such a critical topic as climate change are often subject to personal, political, and media bias—whether for good or for ill.

At its core, however, scientific research is not about politics or 30-second sound bites. Scientific research is about questions and measurable observations. Science is the process of discovery and the means for developing a better understanding of ourselves and the world around us. Science strives for facts and conclusions unencumbered by bias, distortion, and political sensibilities. Although sometimes the methods and motivations are flawed, science attempts to develop a body of knowledge that can guide decision makers, enhance daily life, and lay a foundation to aid future generations.

The relevance and the implications of scientific research are profound, as members of the National Academy of Sciences point out in the 2009 edition of *On Being a Scientist: A Guide to Responsible Conduct in Research*:

Some scientific results directly affect the health and well-being of individuals, as in the case of clinical trials or toxicological studies. Science also is used by policy makers and voters to make informed decisions on such pressing issues as climate change, stem cell research, and the mitigation of natural hazards. . . . And even when scientific results have no immediate applications—as when research reveals new information about the universe or the fundamental constituents of matter—new knowledge speaks to our sense of wonder and paves the way for future advances.

The *Inside Science* series provides students with a sense of the painstaking work that goes into scientific research—whether its focus is microscopic cells cultured in a lab or planets far beyond the solar system. Each book in the series examines how scientists work and where that work leads them. Sometimes, the results are positive. Such was the case for Edwin McClure, a once-active high school senior diagnosed with multiple sclerosis, a degenerative disease that leads to difficulties with coordination, speech, and mobility. Thanks to stem cell therapy, in 2009 a healthier McClure strode across a stage to accept his diploma from Virginia Commonwealth University. In some cases, cutting-edge experimental treatments fail with tragic results. This is what occurred in 1999 when 18-year-old Jesse Gelsinger, born with a rare liver disease, died four days after undergoing a newly developed gene therapy technique. Such failures may temporarily halt research, as happened in the Gelsinger case, to allow for investigation and revision. In this and other instances, however, research resumes, often with renewed determination to find answers and solve problems.

Through clear and vivid narrative, carefully selected anecdotes, and direct quotations each book in the *Inside Science* series reinforces the role of scientific research in advancing knowledge and creating a better world. By developing an understanding of science, the responsibilities of the scientist, and how scientific research affects society, today's students will be better prepared for the critical challenges that await them. As members of the National Academy of Sciences state: "The values on which science is based—including honesty, fairness, collegiality, and openness—serve as guides to action in everyday life as well as in research. These values have helped produce a scientific enterprise of unparalleled usefulness, productivity, and creativity. So long as these values are honored, science—and the society it serves—will prosper."

Important Events in Weapons and Defense Research

1835
French gunsmith Casimir Lefaucheux patents the pinfire cartridge; Samuel Colt patents his design for the revolver.

1883
Maxim patents the first fully automatic machine gun.

1901
The 10-pounder cannon is introduced in Great Britain.

1784
The shrapnel shell is invented.

1855   1885   1915   1945

1882
Armored steel is developed for use in military weaponry.

1916
At the height of World War I, French weapons researchers introduce the modern tank, with revolving caterpillar tracks and a mounted, rotating turret.

1934
The first general-purpose machine gun is introduced.

1936
German company AEG begins developing the first infrared night-vision goggles.

1944
The German Sturmgewehr 44 becomes the first assault rifle to see wide use.

IMPORTANT EVENTS

2012
Flame, a sophisticated data-mining computer virus designed to steal information and possibly wipe out hard drives, is discovered; most affected are personal computers in Iran and other Middle Eastern countries.

1947
The AK47 assault rifle is invented by Mikhail Kalashnikov.

2008
The first littoral (close to shore) combat ship, the USS *Freedom*, is commissioned; the vessel is designed to clear mines, defend against small craft, and hunt for submarines.

1957
Italy develops the 105mm Model 56 pack howitzer.

2006
North Korea test-fires seven missiles and a nuclear weapon.

1950　　1965　　1980　　1995　　2010

1951
The first hydrogen bomb is successfully tested by the United States at Enewetak Atoll in the Marshall Islands.

2007
China carries out a successful test of a ground-based missile that can destroy satellites in orbit.

1995
Lockheed Martin unveils the high-speed stealth F-22 aircraft.

Confronting Threats in the Twenty-First Century

In the early morning hours of May 2, 2011, four American helicopters descended on a high-walled housing compound in Abbottabad, Pakistan. Nearly undetectable on Pakistani radar, two of the choppers hovered while the other two—a CH-47 Chinook and a HH-60 Pave Hawk—moved closer. Far above them, a missile-equipped unmanned drone aircraft circled for extra protection.

Once the HH-60 was in position its door opened and a dozen members of an elite unit—the Navy SEALs—lowered themselves by rope onto a roof of one of the compound's structures. Moments after the last of the SEALs had dropped onto the house the HH-60 suddenly lost power and crashed against a nearby wall; the CH-47 landed safely and more soldiers poured into the compound.

These men had landed in the dead of night to carry out their mission. Intelligence officials in Washington, DC, suspected that the compound sheltered the world's most wanted man: terrorist mastermind Osama bin Laden. For nearly ten years, Bin Laden had remained in hiding after designing, funding, and taking credit for the terrorist attacks of September 11, 2001, in which nearly three thousand people in New York, Pennsylvania, and Washington, DC, were killed. Now, after months of surveillance and planning, the SEAL team hoped to find Bin Laden here and either capture or kill him.

On the ground, the SEALs were well-equipped with the latest weapons available, including short barrel M4 or AR-15 assault rifles loaded with .45 caliber rounds. "The mission dictates the target, the target dictates the weapons and the weapons dictate how they're used,"[1] says former Navy SEAL Richard Machowicz. But before SEAL Team Six could use their rifles they would have to get inside. They blew a hole in one side of one building with explosives and were quickly met with gunfire. They returned fire, killing a woman and two men, and swept through the other compound buildings.

Meanwhile, a smaller group of three SEALs remained in the main house, moving from floor to floor searching for their target. Forty minutes after arriving and sweeping through the entire compound, one of the SEALs spotted a man with a long salt-and-pepper beard and traditional Muslim garb. The SEALs gave chase and cornered Osama bin Laden in a third floor bedroom, along with his young wife. The woman rushed one of the SEALs; she was shot in the leg. Bin Laden himself looked to be unarmed but may have tried to resist. He was shot in the head and chest. His lifeless body was quickly loaded onto one of the helicopters and flown back over the Pakistan border. The manhunt was over.

US soldiers head for a CH-47 Chinook helicopter in Afghanistan. The hunt for the man behind the 9/11 attacks ended with the capture and killing of Osama bin Laden in May 2011. That mission involved Special Forces, intelligence-gathering technology, handheld assault rifles, and aircraft such as the Chinook.

Lives and Missions

Military personnel carried out the mission. But behind them stand the engineers, programmers, aviation specialists, and ballistics experts who spend years designing and developing the tools of their trade—the guns, tanks, aircraft, and other hardware on which their lives and missions depend.

Modern weapons and defense research often builds on, and improves upon, past technology to confront twenty-first-century threats such as terrorist groups and rogue nations. The weapons and defense technologies developed by today's

> **ballistics**
>
> The science or study of the motion of projectiles such as bullets, shells, or bombs.

researchers are also essential tools on battlefields, from Afghanistan to Zimbabwe. Nations such as China, Israel, Russia, France, Great Britain, and the United States employ scientists not only to improve the weapons that already exist but also to develop new, never before conceived defense technology that will give their warriors and military leaders an edge over their enemies. To study weapons and defense is to begin to understand how soldiers in the field fight for their lives and protect themselves as well as how scientists and technicians employ their unique and specialized skills to make the world a safer place.

Weapons experts serve not only armies, navies, and air forces but also intelligence organizations that carry out clandestine missions to observe and spy on adversaries. "The organizations we're talking about have the resources to get any weapon systems they think are necessary to do the job, and they will bring [anything] they think will give them the greatest advantage in that moment," says Machowicz. "If they get it, and they like it, they'll use it."[2] Throughout history, the job of a weapons and defense researcher has been to exploit the latest technology and develop new ideas to provide that advantage.

What Is Weapons and Defense Research?

Science has long been viewed as a driving force behind human progress. From ancient times until today, the curious have worked to understand the processes and structures in nature, the universe, and humanity itself. In the fourth century BC, Greek philosopher Aristotle surmised that parents pass on certain physical and personality traits to their offspring. In February 1953 scientists Francis Crick and James Watson confirmed Aristotle's suspicions when they unlocked the structure of deoxyribonucleic acid, or DNA, which carries hereditary information. In the fifteenth century Polish astronomer Copernicus deduced that the sun, not the earth, is at the center of our galaxy. He was right.

Scientific discovery and research remains a cornerstone of human civilization and a force for progress in the world. "Science," according to microbiologist Cornelius Bernardus van Niel, "is a perpetual search for an intelligent and integrated comprehension of the world we live in."[3] Yet the world can be a dangerous place. One field of science and exploration that remains both essential and controversial is that of weapons and defense research, as experts in computer technology, robotics, biology, chemistry, and other disciplines seek to develop the most potent tools for war and protection against enemy attack.

Warfare and Progress

Warfare is nearly as old as humanity itself. Ancient peoples relied on farming to survive. As populations increased and cities began to grow, the need to protect people and property became a necessary part of life. The early civilizations of Egypt and Mesopotamia rallied around leaders who promised not only to shield their people but also to wage war against those who threatened them. Mesopotamia, the site of present-day Iraq, sat between the Tigris and Euphrates rivers. In nearby Akkad, a warrior named Sargon the Great used his growing military might to conquer and then unite the entire region, from the Persian Gulf to the Mediterranean Sea, in 2300 BC. Sargon's reign inspired the development of sophisticated weaponry and protective armor, including axes, helmets, and horse-drawn chariots.

In the eighteenth century, soldiers fighting in the American Revolution fired slow-to-load flintlock muskets invented a century earlier by French courtier Marin le Bourgeoys for King Louis XIII. During the first modern war, the American Civil War, 1861–1865, the size of the armies mattered less than the technology each side brought to the conflict and hinged on the ingenuity of weapons manufacturers and the mass production of steel for the forging of guns, cannons, and railroads. "The art of war is simple enough," said Union general Ulysses S. Grant. "Find out where your enemy is. Get him as soon as you can. Strike him as hard as you can, and keep moving."[4] While Grant's theory on the art of war is as true today as it was when he said it, the science behind how the enemy is struck, wounded, and defeated has evolved.

Modern Weaponry

The twentieth century witnessed the most dramatic advances in weapons and defense technology the world had ever seen, replacing often unreliable weapons that could be as dangerous to their users as to their intended targets. World War I, fought in Europe between 1914 and 1918, inspired the development of more deadly weapons than ever before. The result was mass slaughter on an unrivaled scale: 17 million deaths—including civilians—and 20 million wounded. One of the most powerful, potent, and versatile weapons developed and perfected during this time was the tank. Its origins stretch back to 1770, when inventor Richard Edgeworth cobbled together a new form of vehicle transport—the caterpillar track, which could be mounted on a tractor's wheels, allowing it to move over muddy and uneven fields. In the succeeding decades, this rotating track combined with steam-powered tractors inspired weapons researchers to consider the vehicle's battlefield potential. In 1899 technician Frederick Simms created what he called a "motor-war car" powered by an internal combustion engine, shielded by a bulletproof glass shell, and armed with two machine guns. Simms's invention was rejected; British military experts saw no practical use for it.

Not until 1915 did a prototype, or trial version, of the tank get serious consideration. In the summer of that year, British army officer Colonel Ernest Swinton and defense secretary Maurice Hankey invited political officials David Lloyd George and Winston Churchill to watch as their primitive, retrofitted tractor sliced through barbed wire during a demonstration. Subsequently, Churchill, who would be Great Britain's prime minister during World War II, ordered further research into the vehicle. Swinton commissioned two men, Lieutenant Walter Wilson and agricultural machinery researcher William Tritton, to develop the land weapon soon known by its code name "tank," so called because of its resemblance to water tanks. Swinton remained involved throughout the research and design processes, insisting that the tank should be able to travel at no less than 4 miles per hour (6.4 kmph) and be able to climb walls and other obstacles as high as 5 feet (1.5m). The tank would also have to bridge five feet (1.5m) of trench and be impervious to rifle fire. Ideally, he told Wilson and Tritton, the vehicle would be manned by a crew of ten soldiers.

> **prototype**
>
> The original or model on which something is based or formed.

In September, Wilson and Tritton rolled out their first version of the tank, called Little Willie. Unfortunately, its slow speed and inability to navigate trenches sent them back to the machine shop. Their second attempt, named Big Willie, fared better, as did their next version, known as Mother, which was 32 feet (9.8m) long and weighed just over 69,000 pounds (31,298kg). On January 29, 1916, Wilson and Tritton rolled their massive, box-shaped tank onto the grounds of London's Hatfield Park. Soldiers cordoned off the area to keep away curious citizens. A second secret test occurred on February 2, and this time British military leaders as well as high-ranking political officials were on hand. Swinton described the experience of seeing his pet project come to life: "It was a striking scene when the signal was given and a species of gigantic cubist steel slug slid out of its lair and proceeded to rear its grey hulk over the bright-yellow clay of the enemy parapet, before the assemblage of Cabinet Ministers . . . and soldiers collected under the trees."[5]

Other trials followed, including one witnessed by King George V. Eventually a lighter model, the Mark I, proved reliable. It was introduced into combat at the First Battle of the Somme in rural France in September

1916. Although the Mark I often broke down and proved noisy and hot for those driving it, British weapons researchers had pioneered one of the most important weapons of the twentieth century.

Rifles and Automatic Weapons

Between World War I and World War II, breakthroughs in weaponry continued. By 1942 the standard issue weapon for most American soldiers was the M-1 Garand Carbine, a gas-operated weapon that held a clip—or bullet storage unit—of eight rounds. Weighing a heavy 9.5 pounds (4.3kg), the rifle was often affixed with a bayonet—or long knife—as long as 16 inches (40.6cm). Developed by Canadian American inventor John Garand, the rifle fired twice as fast as its main competitor, the American 1917 Enfield. The M-1's stock, or exterior, was wooden.

The gun was easy to handle and highly accurate. But like all weapons, it was imperfect. When fired, the M-1 kicked back hard against the soldier's shoulder. Once spent, the clip made an audible "ping" sound, potentially informing those nearby that the enemy's gun was out of ammunition. Despite its flaws, the M-1 served American infantrymen throughout World War II and the Korean War in the 1950s.

Another long-serving weapon that first saw service during the Second World War was the Browning .30-Caliber Air-Cooled Machine Gun. The Browning, though heavy at 41 pounds (18.6kg), could fire up to 550 rounds per minute and, when possible, might be mounted on a jeep. Later modifications allowed it, too, to be used in the Korean War and the Vietnam War in the 1960s.

Also introduced in the years leading up to the Vietnam War was the Colt M-16, perhaps the most versatile and reliable rifle ever made. The US Army ordered eighty-five thousand M-16s in 1963, the gun's first year of service, for the slowly escalating conflict in Southeast Asia. Three years later, the Army asked for another two hundred thousand of the rifles. Makers designed the weapon to be a relatively light 7 pounds (3.2kg) and capable of being fired in one of three modes: cyclical, automatic, or semiautomatic. Although the M-16's ammunition is a relatively small .22-caliber, it can—in cyclical mode—fire up to 950 rounds per minute.

Artillery

The advent of projectile weaponry proved equally ferocious in battle. The pistols and rifles given to soldiers during World War I had advanced little

US troops fire a howitzer during World War I. The development of artillery added a new dimension to warfare by enabling troops to kill or wound large numbers of the enemy simultaneously.

in the preceding decades. Instead, the primary research and development of weapons had been done on artillery, grenades, and mortars—weapons launched into the sky to rain down on enemy troops and kill and maim many of them at one time in massive explosions.

Cannons, the likely descendants of medieval catapults that launched large rocks and other destructive projectiles, were first used in the 1500s. Loaded from the muzzle, or front end, the cannon fired large round shot or cannonballs at targets and caused serious damage to enemy forces. In 1858 a breakthrough in military technology introduced the rifled barrel, which makes the projectile, or shell, spin in flight and more accurately hit its target. Twelve years later, weapons experts developed breech loading—or loading artillery from the back—and fired their shells even further into enemy territory.

In the coming decades came the use of mortars, muzzle-loading weapons that could do more than simply fire shells in a straight line; the shells that came out of mortars were launched in a curved path, guaranteeing better accuracy. One of the more versatile weapons was the British 10-pounder mountain gun used between 1901 and 1915. It first saw action in India during Great Britain's colonial rule of that country. The wheeled gun fired 11.25-pound (5.1kg) shells as far as 18,000 feet (5,486m). Nicknamed the "screw gun" because its long barrel could be broken down into two pieces for easy mobility and then reassembled, the gun was often carried on foot or by mules into the mountains to fire down on enemies. When such weapons were used in trench warfare, their effect was constant and horrific. "Even the rats used to become hysterical," said German lieutenant Stefan Westmann. "They came into our flimsy shelters to seek refuge from the terrific artillery fire."[6] By the end of World War I, the 10-pounder was replaced by the 3.7-inch (9.4cm) pack howitzer, which fired shells that were twice as large as its predecessor.

Artillery remained a vital weapon during World War II, but its place of origin shifted. Although Germany and Sweden had developed their own antiaircraft guns before the war, the introduction of the US 75mm (3-inch) M1A1 marked a turning point in the history of weaponry because of its relatively light weight and compact size. The M1A1 could also launch a 6.24kg (13.76 pounds) high-explosive shell as far as 29,290 feet (8,928m), easily blowing holes in or completely destroying airborne targets. Bigger still was the 105mm (4.13-inch) M3 howitzer, another version of the M1A1.

> **artillery**
>
> Mounted projectile-firing guns or missile launchers, mobile or stationary, light or heavy, as distinguished from small arms.

After World War II Italian gun designers developed the OTO Melara 105/14 Model 56 4.2-inch (10.7cm) howitzer. The Model 56 became the standard artillery in 1957. Despite weighing 3,225 pounds (1,463kg), it could be broken into eleven pieces for easy transport by mule or a cluster of soldiers. Its high-explosive shell could be fired as far as 34,770 feet (10,598m).

Aircraft

Exploding shells and war held no interest for researchers Wilbur and Orville Wright; their interest was manned flight. After years of trial and

The Manhattan Project

The Manhattan Project, one of the most closely guarded weapons projects of all time, began as a race against the clock. In 1939, as World War II began in Europe, noted physicist Albert Einstein sent a letter to American president Franklin Roosevelt warning that the Nazis were attempting to build an atomic bomb. Soon after, Roosevelt ordered the creation of the Manhattan Project, dedicated to completing such a bomb first. Scientists initially had to grapple with enriching, or enhancing, enough of the right kind of uranium, a metallic chemical element, to sustain the kind of chain reaction needed for such a bomb. Scientists used a spinning gas centrifuge to separate the lighter, useful uranium-235 from the heavier and useless U-238.

Finally, on July 16, 1945, in the desert of Los Alamos, New Mexico, the team of scientists led by physicist Robert Oppenheimer tested the device. An enormous white light blazed across the sky, turned orange, and shot upward at 360 feet (110m) per second. At 30,000 feet (9,144m), a massive mushroom-shaped cloud formed. Less than a month later, on August 6, 1945, the US military dropped an atomic bomb—equal to 12,000–15,000 tons (10,886–13,608 metric tons) of TNT—on the Japanese city of Hiroshima in an attempt to force Japan to surrender. Three days later, it dropped a second bomb on the Japanese city of Nagasaki. Nearly two hundred thousand civilians were instantly killed; Japan surrendered soon after.

error, they flew the first powered airplane 20 feet (6.1m) above the beach at Kitty Hawk in North Carolina on December 17, 1903. Soon after the Wright brothers made aviation history, premiers and presidents, infantrymen and generals could envision a time in which war could be waged not only by land or by sea but from the skies. Military aircraft in the decades that followed the Wright Brothers' breakthrough meant superiority for countries that developed fighters and bombers—and terror for armies and civilians that would be attacked from the air.

Before airplanes could be equipped with artillery, fighter pilots often fired pistols and rifles at one another while in midair. At the outset of World War I in 1914, a French Voisin V89 airplane fired its machine gun at a German Aviatik, bringing down the German aircraft—likely the first fighter action of any war in history. Air warfare advanced quickly after that, most notably when Dutch inventor Anthony Fokker developed interrupter gear, which enabled German planes to fire forward through the arc of their propellers. Now pilots could aim their aircraft directly at their enemies. Fokker's achievement was but the first of many. During four years of war, aircraft speeds increased dramatically, from 105 miles per hour (169kmph) to 168 miles per hour (270kmph), and altitudes went from 13,000 feet (3,962m) to nearly 20,000 feet (6,096m).

These early fighters were built from a combination of wood, canvas, and metal, but by the late 1930s all-metal airplanes were the standard. Now, too, machine guns could be mounted on wings, providing more firepower and flexibility. The Messerschmitt Bf-109E could reach speeds of 355 miles per hour (571kmph) in 1939. Five years later, top aircraft speeds were over 428 miles per hour (689kmph).

Aerial Bombardment

World War II began an arms race that many believed would make the difference in who ultimately won the war. "Powerful enemies must be outfought and outproduced," President Franklin D. Roosevelt told Congress in 1941, soon after Japanese war planes attacked the American military base at Pearl Harbor, Hawaii. "It is not enough to turn out just a few more planes, a few more tanks, a few more guns, a few more ships than can be turned out by our enemies. We must outproduce them overwhelmingly."[7]

Japan and Germany may have started the war, but by the spring of 1942 British bombs dropped by the Royal Air Force (RAF) were pummeling German cities. In May, one thousand RAF bombers destroyed 600 acres (243ha) of housing in Cologne, leaving nearly forty thousand Germans homeless. In July the city of Hamburg met a similar fate as British bombs fell by the thousands. The explosions set off massive firestorms that burned to death or asphyxiated another forty thousand people. One German woman remembered the damage inflicted by the RAF aircraft. "Four-story-high blocks of flats (apartment buildings) were like glowing mounds of stone right down to the basement. Everything seemed to have melted and

A British Royal Air Force crew await final checks on their Avro Lancaster bomber, circa 1942. The Lancaster became the most famous and most successful of the World War II night bombers, dropping hundreds of thousands of tons of bombs during thousands of sorties.

pressed the bodies away in front of it. Women and children were so charred as to be unrecognizable."[8]

Jet-propelled aircraft marked another technological leap, beginning in the late 1930s with the German He-178 and continuing with the British Gloster Meteor in 1941. Sleeker and faster than their predecessors, jet fighters have been used in combat operations since the 1950s. "The capacity to win or lose a war actually rested on these weapons' qualities," says weapons expert Chris Bishop, "just as much as it did on the fighting skills of those who employed them and on the strategic sense of those who directed them in their use."[9]

Kalashnikov's Gun

The AK-47 assault rifle, designed by Mikhail Kalashnikov, is likely the most famous and most used handheld weapon in the world. Kalashnikov served in the Soviet army during World War II and was wounded during the Battle of Bryansk in 1941. While recuperating, Kalashnikov began experimenting with the design for a new submachine gun. Soon after, he entered a weapon competition, which required participants to develop a sturdy firearm that could be carried by soldiers in wet and freezing weather. The young soldier's model lost, but he persisted in his work.

The Soviet Army needed an assault rifle to combat their Nazi foes, but the model created in 1944 by weapons designer Alexey Ivanovich Sudayev proved too heavy. Two years later Kalashnikov submitted his own version, and in 1949 the Soviets adopted the 7.62mm (0.3-inch) Kalashnikov assault rifle. Gas-operated, weighing only 11.5 pounds (5.21 kg), and capable of firing six hundred rounds per minute, the rifle has been adopted by gangsters, terrorists, and rebel soldiers around the world. "It's so prevalent because it works," says author Larry Kahaner. "You can drag it through the mud. You can step on it. You can put it under water and it will work every single time. It's very inexpensive, so anybody who feels like they want to start a war can start a war or continue a small war."

Quoted in Andrea Seabrook, "AK-47: The Weapon That Changed the Face of War," NPR, November 26, 2006. www.npr.org.

Biological and Chemical Weapons

Some of the deadliest weapons do not fire bullets, drop bombs, or launch missiles. Instead, these tools of war consist of poisonous chemicals and virulent biological agents. Early in his career at the dawn of the twentieth century, German chemist Fritz Haber focused his scientific research on finding an inexpensive and plentiful alternative to guano, seabird or bat droppings, an effective fertilizer. Guano, in short supply in Europe, had to be imported from Chile, on the other side of the world. The chemical compound in guano—ammonia—was also a primary component in explosives. Working with colleague Carl Bosch, Haber devised a method by which ammonia could be effectively combined with nitrogen and hydrogen. The resulting nitric acid first distilled by Haber and Bosch was a scientific breakthrough that made fertilizer and explosives cheaper and more plentiful than ever before.

Haber's dedication to science was beyond question, but another chemical breakthrough would forever tie him to the suffering and death of thousands. At the outbreak of World War I in 1914, Haber volunteered for service in the German army; he was rejected because he was over forty years old. Instead, the war ministry made him head of its chemical laboratory. In December 1914, Haber and a team of scientists experimented with an artillery shell filled with tear gas, but the gas escaped too quickly. Soon after, Haber filled metal canisters with chlorine, a highly toxic gas. Haber knew that chlorine gas could cause suffocation and death if inhaled. Because chlorine is two-and-a-half times heavier then air, Haber imagined that when released on a battlefield, the gas would snake into the trenches, killing some soldiers and forcing the rest into the open, making them easy targets for German guns. Haber's chlorine canisters were quickly approved by his superiors; he oversaw the production of hundreds of tons of chlorine and thousands of canisters to store it in. Despite Haber's enthusiasm, some in the German military considered the use of poison gas immoral. "I must admit," wrote General Berthold von Deimling, "that the task of poisoning the enemy like rats was repugnant to me, as it would be to any respectable soldier."[10]

Despite the misgivings of some, the first large scale release of chlorine gas occurred at Ypres, France, on April 22, 1915, when 5,730 chlorine canisters were installed along the Western Front. Within ten minutes of release of the gas, French soldiers began choking violently. Within hours, fifteen thousand of them were ailing; five thousand were dead. In early May, distraught over her husband's role in the gas attacks, Haber's wife, Clara, committed suicide with his pistol.

In 1918 Haber was awarded the Nobel Prize in Chemistry for his early work with ammonia, but he left Germany after Adolf Hitler came to power in 1933. He feared his Jewish heritage would put him and his family in danger. Today, Haber is known as the father of chemical weapons. His son Ludwig suggested that German officials saw in his father a driven and dedicated scientist with a limited sense of ethical behavior and little regard for human life: "[The high command] found Haber to be a determined brilliant spirit, an extremely energetic organizer, and a man devoid of scruples."[11]

Napalm

By the time of Haber's escape from his German homeland the use of chemical weapons was more controlled and sophisticated. During World War II, phosphorus grenades and bombs were dropped on targets with the express purpose of setting them on fire. With buildings ablaze, other bombers loaded with explosive bombs could find paths to their targets. The phosphorus itself, the particles of which exploded upon contact with the air, rained down on civilians and soldiers alike, causing scalding skin burns.

> **phosphorus**
>
> A solid, nonmetallic element existing in at least three forms: yellow, poisonous, and flammable; red, less poisonous and less flammable; and black, the least flammable.

Napalm, one of the most fearsome chemical weapons, was invented in 1942 by Harvard University–based chemist Louis Fieser and a team of researchers. Created by mixing aluminum soap powder made of naphthenate and palmitate with gasoline, the jellylike substance, when burned, can reach temperatures surpassing 2,000°F (1,093°C). The extreme heat generated by napalm develops its own high winds that feed upon themselves and create a raging, slow-burning firestorm.

Although used in Japan and Germany during World War II, napalm gained even wider use in Vietnam during the late 1960s and early 1970s. Typically, military personnel dropped the chemical from low-flying helicopters or airplanes and were able to burn jungle foliage, villages, and the people living there with a wall of flames 270 feet (82.3m) long and 75 feet (22.9m) wide. The destruction and death caused by napalm stirred controversy during the Vietnam War, as survivors were left badly scarred by the chemical weapon. An iconic 1972 photograph captures the aftermath of a napalm attack, as young children dash toward the camera. Fieser, though, expressed no remorse for or doubts about his creation: "I have no right to judge the morality of Napalm just because I invented it,"[12] he said.

The Role of Research

Mankind's thirst for war has diminished little in three thousand years, as combatants remain committed not only to self-defense but also to

A napalm strike erupts in a fireball near US troops on patrol in South Vietnam in 1966. Napalm, one of the most fearsome chemical weapons ever developed, gained widest use during the Vietnam War.

conquest. The tools used to fight modern wars, though, have not only transformed the way in which soldiers wage war but also the lives of those caught in the crossfire. Today's fighters carry automatic machine guns and bear down on their enemies in assault helicopters or state-of-the art tanks.

Yet while the traditional weaponry of firearms, tanks, and ships is continually being improved and perfected, the weapons and defense industry has turned many of its resources to areas once considered little more than science fiction. Each year hundreds of companies, employing thousands of workers and spending billions of dollars, research, test, and build new generations of weaponry and defense systems.

Drones

The Arab Spring came to Libya in February 2011. For more than a year, ordinary Arabic-speaking citizens in Tunisia, Egypt, Yemen, Bahrain, and Syria had organized mass protests against their oppressive governments. The antigovernment uprising that began on February 15 in Benghazi, Libya's second largest city, soon spread to the capital of Tripoli. Forces opposing long-ruling Libyan leader Muammar Gadhafi tried to forcefully wrest their country from the dictator, but Gadhafi remained in power, ordering his forces to harshly crack down on protestors and rebels alike.

On March 17 a United Nations Security Council resolution was passed, authorizing a no-fly zone over Libya and any measures necessary to protect the lives of its citizens. Within forty-eight hours, the United Kingdom, France, and the United States began a bombing campaign that they hoped would put a quick end to the worsening Libyan civil war. Instead, the conflict waged for months, but the enormous military effort was aided by the use of highly technical weapons that have now become commonplace: drones.

During the Libyan conflict, these unmanned aircraft flew over the North African country, spying on Gadhafi's forces. Those weeks of surveillance paid off on October 20, when the drones spotted a long convoy of trucks driving near the city of Sirte. Military officials suspected that Gadhafi might be trying to escape. According to an intelligence source, the drones "built up a normal pattern of life picture so that when something unusual happened this morning such as a large group of vehicles gathering together, that came across as highly unusual activity and the decision was taken to follow them and prosecute an attack."[13] That attack consisted of drone-launched Hellfire missiles which destroyed much of the convoy; Gadhafi was indeed attempting a getaway. He survived the missile attacks but was killed soon after by opposition forces on the ground. The drones, a vital part of the effort to oust the dictator, had been remotely piloted, their missiles launched from nearly 7,000 miles (11,265km) away.

Early Unmanned Flight

Throughout the twentieth century, military strategists sought technology that would be not only more effective and accurate in combating enemy threats but would also minimize the danger to military personnel—fighter pilots and infantrymen. Was it possible, researchers wondered, to fly airplanes remotely without having to put people in harm's way? Over the course of nearly a century, the latest technology would be deployed in perfecting the science of unmanned flight. And as the twenty-first century dawned, politicians and generals would find the use of such technology crucial in fighting new enemies around the world. These drones, as they came to be known, would prove deadly as well as controversial.

The earliest known use of unmanned aerial vehicles (UAVs) took place on August 22, 1849. That day, Austrian soldiers launched dozens of unmanned hydrogen-filled balloons packed with explosives and attacked the Italian city of Venice. Only a handful of the balloons worked; the others were blown away by strong winds. More than a decade later, during the American Civil War, tacticians raised explosive-equipped balloons into the skies. The idea was to have them land on enemy ammunition depots and destroy them, but this idea was unsuccessful. What worked better was the use of balloons for reconnaissance, or observation of enemy movement and tactics.

The advent of manned flight in the early 1900s encouraged scientists to experiment with what became the first known warfare drones. Their most daunting challenge was how to control unmanned airborne vehicles, and in the years leading up to World War I, most of them were convinced of the way to do it: by radio. In 1915 the US Navy created the Naval Consulting Board as a way of developing the "machinery and facilities for utilizing the natural inventive genius of Americans to meet the new conditions of warfare."[14] Two of the board's members, inventors Elmer Sperry and Peter Cooper Hewitt, had long been at the forefront of the latest military technology. Sperry and Hewitt, along with Sperry's son Lawrence, developed what they called an aerial torpedo and presented it to a navy representative, Lieutenant T.S. Wilkinson, in the summer of 1916.

Consisting of the Sperrys' patented gyroscope—a mechanical device used to measure and maintain a vehicle's orientation—and Hewitt's radio technology, the aerial torpedo was meant to be installed in an airplane. The plane would then be catapulted into the sky and remotely flown to its target area. There, the controller could have it release a bomb or drop from the sky and crash into its intended target. Wilkinson initially suggested more tests to determine the aerial torpedo's accuracy, but America's entrance into World War I sped up the process. In 1917 the US Navy dedicated $200,000 to Sperry and Hewitt's project. Seven large planes were flown to a field in Copiague, Long Island, for further experimentation.

> **gyroscope**
>
> A device used to help stabilize planes, ships, missiles, and rockets. It consists of a spinning disk or wheel mounted on a base so that its axis can turn in one or more directions and thus maintain its orientation regardless of any movement of the base.

Kettering Bug

The trial and error process started by Sperry and Hewitt led to further development in unmanned vehicle technology, including one developed by Charles Kettering for the US Army. With a range of 40 miles (64.4km), the four-cylinder, forty-horsepower Kettering Aerial Torpedo was built by the Dayton-Wright Airplane Company in Kettering's Ohio hometown. Elmer Sperry built the control and guidance system for what was soon nicknamed the Kettering Bug because of its insect-like appearance. With wings made of cardboard and a fuselage built from a combination of papier-maché and wood laminates, the Bug also featured a carefully calibrated mechanical system. Like its predecessor, the Bug had mixed success, failing more times than it succeeded in test runs. Because of this, officials chose not to deploy any of their forty-five Bugs and kept the weapons program a top secret.

> **fuselage**
>
> The complete central structure to which the wing, tail surfaces, and engines are attached on an airplane.

In the years after World War I, the American military continued its drone tests. So did its close ally, Great Britain. Between 1927 and 1929 the Royal Navy tested an airborne vehicle named the Larynx. Constructed from a small plane that could be launched from a warship and flown on autopilot, the Larynx met with modest success and led to further development of radio-controlled aircraft.

Other models soon followed, including the British Fairey Queen in 1931 and the DH.82B Queen Bee in 1935. The Bee's mechanical descendants soon became known as "drones," named for the nonstinging male honey bees that develop from unfertilized eggs in the hive. Drones are known for their allegiance to the queen bee and their speed of flight.

Hitler's Drones

Speed and allegiance were the very qualities that German führer Adolf Hitler sought in his drone program. The Nazi leader dreamed of dominating the European continent in the 1930s and ordered German scientists to begin work on a flying bomb that he planned to use on civilian targets. Engineers Robert Lusser of the Fieseler company and Fritz Gosslau from the Argus engine works soon designed the Fieseler Fi-103, also known as the *Vergeltungswaffe* (revenge weapon)-1, or V-1. Launched from a long ramp, the V-1 had a thrust pulse-jet gave it a loud buzzing sound, and it could reach speeds up to 470 miles per hour (756kmph).

Preprogrammed to fly approximately 150 miles (241km) and crash into its target, the V-1 was first launched against Great Britain in 1944 where it killed at least nine hundred civilians and injured thousands more. Those on the ground would first hear the buzzing sound made by the V-1s—their signal to take cover. Then, suddenly, the sound would cease. Moments would pass and then the V-1 would come crashing to the ground and explode in a massive fireball. "The suspense was appalling," says historian Taylor Downing. "The flying bombs could come over at day or night and regardless of the weather. They were impersonal, indiscriminate killers and made people feel helpless."[15]

The devastating effects of Hitler's V-1s prompted the US Navy to redouble its own UAV program. The goal was to build drones that could attack V-1 launch sites. In 1944 commanders ordered the navy's Special Air Unit 1 (SAU-1) to transform four-engine B-17 bombers, nicknamed "Flying Fortresses," and other planes into unmanned vehicles that could be packed with 25,000 pounds (11,340kg) of explosives and remotely controlled using television guidance systems. The plan of assault was simple but dangerous: the planes would be manned by two pilots who would fly the aircraft to a low altitude—no more than 2,000 feet (609m)—before programming the plane to hit its target and parachuting to safety. Though late in the war, this method succeeded in destroying dozens of V-1s before they could be launched. Never before had drones destroyed drones.

The RQ-1 Predator drone (pictured) can fly over dangerous areas and transmit video, infrared, and radar information to controllers on the ground. It has been used by both the US Air Force and the CIA.

Advances in Drone Technology

In the years after World War II, drone research lagged while military scientists, especially in the United States, turned to the development of rockets and missiles. Not until the early 1960s did air force engineers return to UAVs. When they did, surveillance drones became their focus. But to equip any aircraft to study the enemy, stealth technology that would allow the drone to go undetected was essential.

Therefore, the US Air Force engineers modified combat drones—particularly the Q-2C Firebee—for new missions by placing a screen over the vehicle's engine air intake and packing special blankets between the plane's fuselage and the interior. In this way they were able to blunt the drone's radar signal by which enemies could track it and shoot it out of the sky. As a final stealth technique, workers slathered the aircraft with a special antiradar paint. The newly named AQM-34 Firebee was launched from a DC-130 director plane. Once its mission

The Ethics of Drone Warfare

Drone attacks by the United States against terrorist threats have greatly increased under the administration of President Barack Obama. By the beginning of Obama's third year in office, drone strikes had killed more than fifteen hundred militants in Afghanistan, Yemen, Pakistan, and Somalia compared to less than fifty killed under his predecessor George W. Bush.

Warfare experts have questioned the legal and ethical use of drones on a variety of grounds. Although American military pilots typically fly drones, they are often authorized by the CIA, a civilian agency which does not have to abide by the standard rules of combat, thus complicating international treaties and agreements. The ethics of drone use are further complicated when non-terrorists are killed by mistake. Estimates suggest that hundreds of civilians have died in recent years after being mistaken for militants or being at the wrong place at the wrong time.

Still, few deny the effectiveness of drones in combating terrorists around the world. In 2012 Obama strongly defended the use of drone warfare as a focused and accurate way of killing those who seek to harm Americans, particularly members of the terrorist group al Qaeda. "We have been very careful about how it's been applied," Obama said. "Al Qaeda's been really weakened, but we've still got a little more work to do, and we've got to make sure that we're using all of our capacities to deal with it."

Quoted in Carol E. Lee and Adam Entous, "Obama Defends Drone Use," *Wall Street Journal*, January 31, 2012. http://online.wsj.com.

was completed, the AQM-34 deployed a parachute, which enabled it to land safely for pickup by a helicopter.

The AQM-34 was used throughout the Vietnam War in Southeast Asia and during the Cold War against targets in the Soviet Union. So successful was the American UAV program that other nations began their own experiments with drones. Israel, in particular, believed that it could utilize UAVs against its enemies in Egypt, Syria, and Lebanon. In 1970 the Jewish state bought twelve Firebees from the United States, modified their design and flying capabilities, and renamed them the Firebee 1241s. Less than three years later the Israeli Air Force used its new drones in the Yom Kippur War. On the second day of the conflict, the Firebees led the attack against Egyptian air defenses in the Suez. In only a matter of hours, the Egyptians had expended their entire stockpile of surface-to-air missiles—forty-three in all—trying to bring down the Firebees. The Firebees survived the onslaught and successfully destroyed eleven of the Egyptian missiles.

Over the next two decades, Israel developed UAVs of its own, eventually becoming a leader in the industry. Israel Aircraft Industries (IAI) built the Scout, which had a 13-foot (4m) wingspan. Its fiberglass frame blunted the craft's radar signature, making it hard to detect and destroy. A video camera mounted in the center of the Scout transmitted real-time surveillance data.

New Generation

Today, advances in computer technology and aviation have ushered in a new generation of drones, also known as Remotely Piloted Aircraft, or RPAs. Combat drones can be launched against single targets, including terrorists, or they can be utilized as one part of a larger combat mission, launching missiles and retrieving intelligence. Drones are also cost effective. On average, a drone costs $10.5 million to produce; jet fighters cost $150 million. The remotely located drone pilot controls the aircraft while watching a live video feed with a two-second delay. Intelligence coordinators double-check locations using Google Earth maps. In preparation for a missile strike, a sensor operator helps the pilots use a laser beam to guide the missiles to their targets. Commanders, pilots, and coordinators can communicate with one another during operations via secure Internet chat rooms.

One of the most widely used armed drones is the RQ-1 Predator. First seen during American air strikes in the Balkans in the 1990s, Predators have also been used against targets throughout the Middle East and Asia. Designed and engineered by General Atomics Aeronautical Systems, the Predator is used by both the US Air Force, which has more than sixty of them in its arsenal, and the Central Intelligence Agency (CIA). Initially designed for reconnaissance, the Predator is now typically equipped with high definition cameras so that ground personnel can observe potential targets—trucks, buildings, people—before launching an attack. For that purpose, the Predator is loaded with Hellfire antitank missiles. Controllers maneuver the drone by either a line-of-sight radio connection or a satellite hookup.

Northrop Grumman's jet-powered Global Hawk RPA has a wingspan of 131 feet (39.9m) and can reach an altitude of 65,000 feet (19,812m)—twice the cruising altitude of most commercial airplanes. The Hawk can explore thousands of square miles a day for the purpose of relaying images to military headquarters. Depending on the mission, a Global Hawk could be aided by the RQ-9 Reaper. The Reaper has the advantage of being able to fly for more than twenty-four hours, and technicians can program it to patrol specific areas with no need for a controller.

Ongoing Challenges

Still, while many consider combat drones a superior weapon for fighting modern wars, remote-controlled weaponry remains an imperfect technology. In the fall of 2011, the American military's security system discovered a computer virus that logged, or kept track of, every computer keystroke made by drone pilots. That data, much of it detailing drone locations as well as targeting information, is meant to be classified. Officials fear that this secret data may be transmitted by America's enemies over the Internet and used to foil drone attacks. Despite technicians' best efforts to wipe out the virus, the infection only returns. "We keep wiping it off, and it keeps coming back," says one anonymous military source. "We think it's benign. But we just don't know."[16]

Another concern for drone researchers and security personnel is the failure to encrypt, or make unreadable, the video transmitted by drones. Encryption masks top secret information that could fall into the hands of terrorists or enemy nations. In 2009 American soldiers discovered hours of drone video on the laptop computers of insurgents in Iraq, signaling that drone strategy was being carefully studied by America's enemies.

> **encryption**
>
> To put into, or change, secret computer code so that others cannot understand or read it.

Nuclear-Powered Drones

Concerns over drone technology extend to plans for nuclear-powered drones. Imagined by researchers and designers as early as the 1950s, nuclear powered aircraft is only now a viable technological reality. The blueprints for the drones—developed by US nuclear research agency Sandia National Laboratories and Northrop Grumman, a defense contractor—could transform the industry.

With a nuclear-based power source, the drones could remain in the air for months without refueling. Reaper drones carry 2 tons (1.8 metric tons) of fuel and an equal weight in missiles and other equipment; they can remain in the air for forty-two hours with light payloads and fourteen hours fully loaded. A nuclear-powered Reaper could carry more missiles and munitions and remain in the air for weeks longer with no need for ground crews to refuel and reequip the aircraft. Communications and the ability to run more complex weapons systems would also be greatly enhanced.

Despite their obvious advantages, these "ultra-persistence technologies," as researchers refer to them, have raised debate. Because the drones could crash, experts fear that a drone disaster could expose people to the deadly nuclear material carried by the crafts. "It's [a] pretty terrifying prospect," says Chris Coles of Drone Wars UK, which campaigns against the increased use of drone technology. "Drones are much less safe than other aircraft and tend to crash a lot. There is a major push by this industry to increase the use of drones and both the public and government are struggling to keep up with the implications."[17]

In April 2012 those implications halted the development of nuclear-powered drones. After years of research, Sandia and Northrup Grumman were ordered by the US Department of Defense to stop their work on the technology, at least for now. While these sensitive weapons systems and the decisions made about them are typically kept secret from the public, military experts suggest that political pressure may have been involved. The public outcry over the safety of nuclear-equipped aircraft might have been too great.

Drone Controversy

Contemporary drones have not escaped controversy either. For some, waging war at a distance is part of the problem. In 2011, twenty-three-year-old navy hospital worker Benjamin David Rast and twenty-six-year-old marine staff sergeant Jeremy D. Smith were killed in Afghanistan when a Predator drone mistakenly fired on them. The unnamed captain who pushed the button to launch the missile did so from Creech Air Force base near Las Vegas, Nevada, 7,500 miles (12,070km) away. The captain believed the young Americans to be Afghan insurgents.

Incidents of "friendly fire" have occurred in virtually all wars. Still, Rast's father, Robert Rast, has been critical of the military and wants those involved in his son's death to be held accountable. "The Predator crew responsible for this particular airstrike not only destroyed a young Navy corpsman's life but also destroyed the lives of his entire family,"[18] Rast wrote in an e-mail to the *Las Vegas Review-Journal*. He also criticized the ease with which such attacks are launched, comparing them to playing video games. Rast believes that firing on targets from so far away has become too easy and increases the number of mistakes made by pilots.

Pakistanis protest US drone strikes on terrorist hideouts in their country. The increasing use of drones for conducting missile strikes has prompted angry demonstrations in Pakistan, where some of these strikes have killed innocent civilians.

Drones Close to Home

Although drones were originally developed for military use, in February 2012 the US Congress passed a bill to allow them in civilian airspace, along with private and commercial planes. Officials predicted that by the end of the decade, as many as thirty thousand drones could be flying over the United States. A majority of those drones would be used by federal and local law enforcement to track criminals, hunt domestic terrorists, and provide data on the weather and the atmosphere. Other drones could be used commercially in businesses ranging from surveying to real estate. One real estate agent, Daniel Gárate, already employs a small, remote-controlled drone to photograph exclusive properties that he hopes to sell, providing potential buyers with breathtaking aerial views of homes.

The widening use of domestic drones is not without controversy either. Aside from the fear of overcrowding in the skies, many believe that drones intrude on the privacy of average citizens. A statement by the American Civil Liberties Union (ACLU) expressed concern for the privacy of American citizens if wider drone use were approved: "Unfortunately, nothing in the bill would address the very serious privacy issues raised by drone aircraft. This bill would push the nation willy-nilly toward an era of aerial surveillance without any steps to protect the traditional privacy that Americans have always enjoyed and expected."

Quoted in MSNBC, "Pilots Worry About Safety of Allowing Domestic Drones in US Skies," February 7, 2012. http://usnews.msnbc.msn.com.

The distance between the drone pilots and the drones themselves therefore remains a controversial aspect of drone warfare.

By 2012 drones were also at the center of controversy between the United States and Pakistan. Between 2005 and May 2012 the United States carried out 251 drone strikes against suspected terrorists in Pakistan. Some of those attacks killed innocent civilians. Pakistanis have expressed outrage at the use of drones, which they believe are more likely to lead to wrongful deaths than traditional piloted aircraft. "The drones are killing innocent bystanders, including children and women," says Chaudhry Nisar Ali Khan, a member of Pakistan's parliament. "They

Drones of the Future

Researchers are working on the development of drones that might one day be able to navigate without a real-time human controller—meaning they could hunt, target, and kill enemy forces autonomously. Advances in weapons technology, particularly the creation of weapons systems capable of lethal autonomy, have raised questions about whether or how such technology should be used.

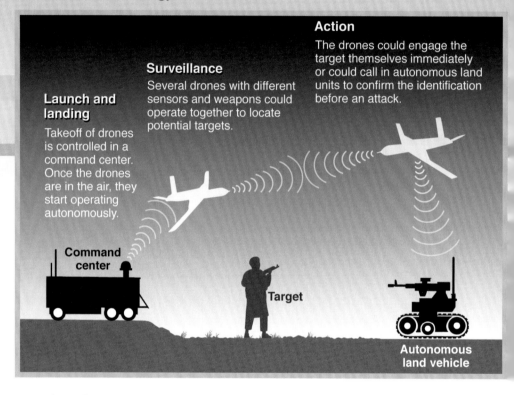

Source: Georgia Tech Research Institute. Alberto Guadra and Peter Finn/*Washington Post*, September 19, 2011.

must be stopped forthwith."[19] But the success of the drone attacks means the military is unlikely to stop using them. Despite civilian casualties, the drones remain precise and powerful when used against militants who hide in Pakistan's mountainous region.

Drone Competitors

The success and relatively low cost of American drone warfare has inspired many other countries to begin their own UAV projects. In 2006

China introduced its first drone. By 2021 the Asian nation will have spent $94 billion to produce a fleet of UAVs. In late 2010 the Chinese government unveiled twenty-five prototype UAVs at an air show. Designed and built by the researchers and scientists at ASN Technology Group, Chinese Aerospace Science and Technology Corporation, and China Aerospace Science and Industry Corporation, some of the models may be able to fly faster than Predator and Reaper drones.

India, Russia, and Pakistan are also developing drones, although when these UAVS will be operational remains unclear. Drones are likely to remain an indispensable weapon of war for the foreseeable future. They may one day even replace manned air flight altogether. "This is the direction all aviation is going," says law professor Kenneth Anderson, who studies the legality of drone warfare. "Everybody will wind up using this technology because it's going to become the standard for many, many applications of what are now manned aircraft."[20]

Over the course of the twentieth century, UAVs evolved from rickety, unreliable vehicles into sleek, dependable machines that can transmit high-quality photographs and video as well as carry out deadly missions with precision. Their wider use and development around the world will likely bring about a host of new advances that scientists are already dreaming about.

Cyberwarfare

In July 2010 a computer technician working for a security company in Belarus in Eastern Europe noticed something odd on his screen. A worm, or coded computer virus, appeared to be affecting the software of one of the company's clients—the Islamic Republic of Iran. Initially, investigators believed that the worm had been created to steal Iranian nuclear secrets. For close to ten years, the Middle Eastern country had been attempting to develop nuclear power; other countries feared that despite claims to the contrary, Iran was also planning to build nuclear weapons.

Upon further analysis, researchers discovered that the worm, soon named Stuxnet, attacked software systems for the purpose of sabotaging—in this case slowing down—the emerging Bushehr nuclear reactor's computer network. Engineered into Stuxnet was the ability to detect particular software settings, find vulnerable spots in those settings, and inject its own disruptive code into the system. "It then spreads from machine to machine on the network," says computer worm expert Patrick Fitzgerald, "exploiting a second vulnerability to do so, and reports back to the attacker."[21] The worm was searching for a particular type of software that would allow it to disrupt the Iranian system.

Delays in the construction of Bushehr occurred at a time that suggests Stuxnet may have been to blame. Articles in the *New York Times* identify the United States and Israel as creators of the virus but neither country has admitted any wrongdoing. Computer-spread worms or viruses such as Stuxnet are, in the words of some experts, the newest frontier in modern warfare. The cyberworld will become "more important in the conflict between nations," says Israeli deputy prime minister Dan Meridor. "It is a new battleground, if you like, not with guns but with something else."[22]

Growing Concern

That new and dangerous battleground of cyberattacks and cyberwarfare is used by nations and terrorists alike. The twenty-first-century world is linked by complex networks of computers. This interconnectedness makes many aspects of daily life run smoothly: the electricity that lights

CHAPTER THREE

a house, the system that provides clean and safe drinking water, and the traffic lights that keep traffic moving are but three of the systems controlled by computers.

The intentional disabling, disruption, or destruction by computer hackers of this infrastructure may constitute the beginning of a cyberwar. This kind of politically motivated hacking, or electronic breaking and entering, is more common than ever before and potentially poses a threat nearly as lethal as conventional weapons. For that reason, countries and companies around the world are spending millions of dollars to develop their own cyberweapons and stronger cyberdefenses.

> **infrastructure**
>
> The basic, underlying framework or features of a system or organization.

In 2009 Barack Obama announced the creation of the US Cyber Command (USCYBERCOM). Headquartered at Fort Meade, Maryland, USCYBERCOM develops new cyberresources, including weaponry and defense systems. In his remarks, Obama spoke of the seriousness of the dangers facing the country. "It is now clear this cyber threat is one [of] the most serious economic and national security challenges we face as a nation," he said. "We know that cyber intruders have probed our electrical grid, and that in other countries cyber attacks have plunged entire cities into darkness."[23] Obama also signed an executive order detailing the strategies military commanders can use to launch cyberattacks. The guidelines, developed over two years by the Pentagon, include information ranging from how to plant computer viruses to how best to cripple an enemy's electrical grid or defense network.

Tools of the Trade

Today, both private contractors and government agencies alike are secretly developing a variety of cyberweapons and defenses against cyberattacks. Some of this research focuses on finding vulnerabilities in the computer systems of other countries. Government security expert Richard A. Clarke defines cyberwarfare this way: "Actions by a nation-state to penetrate another nation's computers or networks for the purposes of causing damage or disruption."[24] Damage, though, does not necessarily mean destruction. Internet espionage is of particular concern to cyberexperts. Working from hard-to-pinpoint locations, faceless hackers who work to find a breach in Internet security, infiltrate a company's or government's

President Barack Obama created the United States Cyber Command, whose logo appears here, to develop new cyberweapons and cyberdefense systems. The federal government views cyberattacks and cyberwarfare as serious threats to national security.

network firewall, a shielding device intended to protect classified or proprietary information not meant to be seen by the general public.

One tool employed for the purpose of cyberespionage is a rootkit. A rootkit is a type of software that can be integrated into a computer and remain hidden without the knowledge of the user. Like tiny, internal spies, rootkits are not made to destroy a computer or network but to take note of passwords and sometimes to modify programs stored on the machine. Equally insidious is a bot, or automated computer program, for distributing viruses. When linked with hundreds, thousands, or millions of other computers, these bots form a botnet, or network. Once connected, the bots are marshaled by a "botmaster," or "bot herder," who can use the botnet to clog an Internet site, rendering it unusable or vulnerable to spying. "If you were to launch with a botnet that has 10 million computers in it—launch a denial of service attack—you could launch a large enough attack that it would not just overwhelm the target of the attack, but the root servers of the Internet itself, and could crash the entire Internet," says journalist Mark Bowden. "What frightens security folks, and increasingly government and Pentagon officials, is that a botnet of that size could also be used as a weapon."[25]

> **rootkit**
>
> A type of software that can be integrated into a computer and remain hidden without the knowledge of the user.

Once inside the target's computer network, the cyberspy may attempt to steal secrets. At one time, nations depended on human spies to obtain such information; today, sensitive secrets can be plucked by a single hacker from the other side of the world. One of the earliest and most notorious incidents of cyberespionage was uncovered in 1999. Moonlight Maze, as it became known, occurred when American officials detected that a variety of computer systems at NASA, the US Department of Energy, the Pentagon, and a number of research labs and universities were being probed by a foreign source. Thousands of critical military documents were being viewed, likely copied, and perhaps used by the mysterious hacker or the people he or she worked for. After an exhaustive investigation, the US Department of Defense followed the cybertrail to a computer in the former Soviet Union, America's archenemy during the Cold War of the 1950s and 1960s. The Russian government denied knowledge of the invasion.

Enemies and friends alike use cyberespionage to "listen in" on their international counterparts. In early 2012 American ally India was accused of eavesdropping on a US commission's e-mails relating to economic and security concerns between China and the United States. Investigators traced the Internet infiltration to an Indian government spy organization interested in the relationship between the two nations. But the exact source of the espionage and what the hackers hoped to gain from the information remains unclear. "There is some malicious intent, but to try and work out who has done it, given the current nature of the Internet, is an exercise in futility,"[26] says Cherian Samuel, a specialist on cybersecurity.

Threats to National Security

Government secrets can create tensions between nations, but cyberespionage affects average citizens even more directly and can threaten a country's national security. When a company's computer networks are assaulted by Internet thieves, a client or user's personal information is typically the target. For years people throughout the world have been vulnerable to identity theft, in which enterprising criminals obtain a credit card number or other information that they can exploit to steal money. Millions of people use the Internet to buy products, and although online companies promise that buyers' private information is safe and secure, hackers work hard to break through security systems to get at this valuable information. According to some experts, tens of thousands of attempted hackings per day occur in the United States alone. In 2009 American banks lost a combined $100 million through illegal cyberinvasions. According to a report by the US National Security Council, this kind of theft affects institutions and individuals. "Through cyber crime, transnational criminal organizations pose a significant threat to financial and trust systems . . . on which the world economy depends," the report states. "Pervasive criminal activity in cyberspace not only directly affects its victims, but can imperil citizens' and businesses' faith in these digital systems, which are critical to our society and economy."[27]

> **botnet**
>
> A group of computers that are controlled from a single source and run related software programs.

In April 2011 the Sony Corporation, maker of the popular PlayStation, announced that the personal information of its 70 million network users had been exposed by hackers. Aside from names, addresses, and passwords, Sony suggested that the credit card information of its users might also have been taken. Three months later the social networking site of SK Communications, one of the largest companies in South Korea, was hacked. Perpetrators pilfered the names, home and e-mail addresses, and other identifying information of as many as 35 million SK clients. The prime target for such hackers is the "endpoint machine," or a person's home or work computer, says cybersecurity expert Alastair MacGibbon:

> It basically has value to a criminal both for the information that's on it, so for example your personal details, maybe your passwords and

other things that a criminal can either use themselves or trade with others. And then the machine itself can have value. So I can store my documents or other criminal things on your computer rather than on my own. Or I can use that computer with a series of other computers . . . to actually launch attacks against other systems.[28]

In March 2012 Visa and Mastercard, America's two largest payment processors, announced breaches in their security systems. Specifically, officials said, the credit and debit card information of 1 million to 3

Hacker Informant

In early 2012 an infamous hacker calling himself The Real Sabu dropped an Internet bombshell. For months he had encouraged his followers in the hacking community to attack government agencies and private companies by sabotaging and breaking into computer networks around the world. Sabu and these other saboteurs are part of a loosely organized, politically active global group known as Anonymous. Sabu, whose real name is Hector Xavier Monsegur, was first arrested in New York in August 2011 and eventually pleaded guilty to twelve counts of conspiracy to attack computers before returning to his hacking ways.

What none of Sabu's fellow hackers knew is that their leader had turned government informant, helping to disrupt, track down, and arrest them. Five men from Great Britain, Ireland, and the United States were arrested in March 2012 and charged as computer criminals based on information provided by Sabu. Government officials remain hopeful that by infiltrating the ranks of groups like Anonymous, they can turn hacker against hacker and stem the tide of Internet sabotage. "It is going to be very difficult for Anonymous to recover from such a breach of trust," says Mikko Hypponen, a security researcher in Finland. "You can see the Anonymous people now looking left and right and realizing, if they couldn't trust Sabu, who can they trust?"

Somini Sengupta, "Arrests Sow Mistrust Inside a Clan of Hackers," *New York Times*, March 6, 2012. www.nytimes.com.

million people had been exposed. Although the companies contacted these cardholders, the incident provided another example of how Internet criminals attack. Ominously, spokesmen for Visa and Mastercard were unclear as to how the breaches in their systems occurred, making future attacks likely.

> **root server**
>
> Any of the domain name system servers on the Internet that contain the IP addresses of the top-level domain registry organizations that maintain global domains such as .com, .org, .net, and country code domains such as .uk and .ca.

In most cases, the security software used by corporations is simply not sensitive or strong enough to repel attacks. "Hackers are well aware that these systems don't have the same sophisticated levels of security as the banks,"[29] says Tom Kellerman, a computer security expert. More unsettling is that the techniques used by hackers are often so subtle, so difficult to detect, that security breaches take months, even years, to discover. An American company, Heartland Payment Systems, was hacked in 2007, but officials did not discover it until 2009. During that time, at least 130 million consumer credit card numbers were taken, ultimately costing Heartland $140 million dollars in legal fees and financial settlements with customers. On average, 416 days will pass before a company detects that its networks have been violated.

Sabotaging the System

Internet thievery and spying is a serious threat to any nation and its citizens but, perhaps, even more consequential is sabotage, where hackers purposely try to demolish or interrupt a computer network. Armed with little more than a hard drive and miles of computer code, their goal is to create mischief, even devastation, on a large scale. Of most concern to law enforcement is a hacker's ability to shut down or cripple the complex computer systems that control a country or city's infrastructure: water, electric, or gas.

"If I were an attacker and I wanted to do strategic damage to the United States," says retired admiral Mike McConnell, "I probably would sack electric power on the U.S. East Coast, maybe the West Coast, and attempt to cause a cascading effect. All of those things are in the art of the possible from a sophisticated attacker."[30] Such interruption of the power grid would not only keep people in the dark. It could leave them

The possibility of cyberattacks against the infrastructure that keeps the country running—electrical power, water, and gas—represents a new threat and a new challenge for researchers and cybersecurity specialists.

freezing in the winter and sweating in the summer heat. Illness and even death would likely result and leave federal and local governments at the mercy of countries or individuals intent on doing harm to large numbers of people around the world.

Many nations, including the United States, now consider this kind of wanton Internet destruction an act of war, which could be met with

Digital World War

While most computer viruses are aimed at particular governmental or commercial networks, the Conficker worm attacks the Internet itself. The worm is a self-updating malware that first burrows its way into a computer and operates without the permission or knowledge of the computer's owner. "What Conficker does is penetrate the core of the [operating system] of the computer and essentially turn over control of your computer to a remote controller," says writer Mark Bowden. "[That person] could then utilize all of these computers, including yours, that are connected. . . . And you have effectively the largest, most powerful computer in the world."

First discovered at Stanford University in 2008, Conficker prompted a group of volunteer experts calling itself the Conficker Working Group to notify the American government, but officials at the time had little understanding of cyberwarfare. After close study, the Conficker Working Group realized that the unknown people behind Conficker had little interest in disrupting or destroying the Internet. Instead, they wanted to make money by stealing passwords and siphoning millions of dollars from the world's banks. Thus far, more than 12 million computers have been infected with Conficker worldwide and cybersecurity experts remain on high alert, hoping to defend against Conficker and its mysterious makers if and when they decide to cripple the world's computer networks, compromise the Internet, and wage an all-out digital world war.

Quoted in NPR, "The 'Worm' That Could Bring Down the Internet," September 27, 2011. www.npr.org.

---

military retaliation. "A cyber attack," says retired air force major general Charles Dunlap, "is governed by basically the same rules as any other kind of attack if the effects of it are essentially the same."[31] Cybersabotage can have disastrous consequences, and nations are actively working to not only defend themselves but, if necessary, to launch counterattacks to incapacitate their enemies with computer software. Although most will never admit to the practice, these countries consider cyberwarfare a fact of life in the twenty-first century.

Fighting a Cyberwar

What constitutes an act of war has become foggy and unclear, as have the rules of engagement by which countries or organizations wage a conflict. Is there a difference between a cyberwar and a cyberattack, which happens in nearly all industrialized countries every day of the year? Once, the sides were clearly identified, but in the tech age, this has changed. Where armies once waved their flags proudly as they marched forward into battle, cyberwar is launched from the shadows. Who is sabotaging whom often remains hidden, and the evidence against a perpetrator is usually circumstantial. "One thing about war is that, historically, the lines have been drawn and there is an understanding of who the enemy is," says David M. Nicol of the Information Trust Institute at the University of Illinois at Urbana-Champaign. "When a cyberattack occurs against a sovereign state, who do you declare war on?"[32]

The cyberscientist's goal is to hamper as much as possible the saboteur's ability to wreak havoc. But enhanced security software may not be enough. "We've been playing defense for a long time," says former FBI executive assistant director Shawn Henry. "You can only build a fence so high, and what we've found is that the offense outpaces the defense, and the offense is better than the defense."[33]

In cyberspace, pretending to be someone else is common. This only complicates law enforcement's ability to capture hackers. Once a hacker's location or country is determined, military action can be taken, but proof, or lack thereof, can be the difference between war and peace. "Clearly," says former Pentagon cyberexpert O. Sami Saydjari, "you want to be able to attribute an attack with a degree of certainty before you respond with military action."[34]

Although it is difficult to prove, cyberexperts suspect that the countries most likely involved in cyberattacks are China, Russia, and the United States because they have the financial and technological resources necessary to carry them out. Experts say that tracking down the origination point of cyberattacks is the biggest challenge they face, and according to some, that cyberbattle is being lost. Computer criminals are too good at what they do; cybersecurity is simply not good enough.

"I don't see how we ever come out of this without changes in technology or changes in behavior," says Henry, "because with the status quo, it's an unsustainable model. Unsustainable in that you never get ahead, never become secure, never have a reasonable expectation of privacy or security."[35] Beyond building stronger firewalls and other security software and protective defenses, more officials are calling for better international agreements and more transparency, so each side can see what cybertools the other has in its computer arsenal. But many countries appear unwilling to cooperate.

Chinese Cyberattacks

For years, western nations have been dismayed and angered by what they believe to be cyberattacks conducted by the Chinese government in Beijing. In April 2012 mutual distrust between the United States and China turned to something like cooperation when officials from both countries revealed their recent participation in a series of war games. Meant to defuse any potential military escalation, these cyberscenarios mimic how each side might behave if one waged a cyberattack on the other.

The make-believe scenarios are treated seriously, though informally, as officials from both sides make decisions and give orders about protecting infrastructure computer networks and state secrets. The goal is to open a dialogue between the two countries and to resolve potential conflicts. "The officials start out as observers and become participants," says Jim Lewis, director at the Center for Strategic & International Studies (CSIS) in Washington, DC. "It is very much the same on the Chinese side. Because it is organised between two thinktanks they can speak more freely."[36]

Whether the war games can improve the United States–China relationship remains to be seen. Both the United States and its key ally the United Kingdom have long suspected China of stealing billions of dollars worth of plans from defense contractors and government agencies, most of it through the use of hackers. China claims to stand firm against cybercrimes. "It is hard to attribute the real source of attacks," says Chinese defense minister Liang Guanglie, "and we need to work together to make sure that this security problem won't be a problem."[37]

Until international agreements can be reached, all sides will employ scientists to strengthen their cyberdefenses. In the United States alone, spending for cyberdefenses has more than doubled, from $88 million in 2009 to $208 million in 2012. The price of protection will remain high for the foreseeable future. "Once the province of nations, the ability to destroy via cyber now also rests in the hands of small groups and individuals," says William Lynn, US deputy defense secretary. "From terrorist groups to organised crime, hackers to industrial spies to foreign intelligence services. . . . This is not some future threat. The cyber threat is here today, it is here now."[38]

Protecting Troops

CHAPTER FOUR

Like so many children, young Stephanie Kwolek looked to her parents for guidance and inspiration. Born in 1923 in New Kensington, Pennsylvania, Kwolek often walked the woods with her father, a naturalist, looking for birds and identifying the trees and flowers they found there. She would collect seeds, grasses, and wildflowers to take home and press into scrapbooks. Kwolek's idyllic world collapsed when her father died suddenly; she was ten years old. Her homemaker mother, who shared her love of fabrics and sewing with her daughter, had no choice but to find work to support the family.

As she grew, Kwolek retained her love of nature and found it best expressed in the fields of medicine and chemistry. After graduating from Margaret Morrison Carnegie College, Kwolek found herself at a crossroads. "When I got out, I realized I didn't really have enough money to go to medical school," she says, "so I decided to work in the field of chemistry and save my money with the intention of eventually going to medical school."[39] Kwolek applied for the position of chemist at the DuPont Company, was offered a job, and took it. She never got to medical school.

Instead, working in polymer research, the ambitious and hardworking chemist made a number of scientific advances in developing chemical fibers that could perform in and withstand extreme conditions. Then, in 1965 Kwolek discovered that certain heavy, rod-like molecules created in the laboratory could be woven together to create a stiff, nearly impenetrable fabric. When mixed together, the solution looked like buttermilk, rich and creamy. Kwolek's unique chemical recipe—soon after dubbed Kevlar—would in a few short years be used in the construction of tires, boats, tennis rackets, skis, and airplanes.

In the early 1980s the US military began weaving vests made of Kevlar because they found that the chemically dense fabric was virtually bulletproof. Today, Kevlar-based body armor is standard issue equipment for soldiers all over the world. Kwolek's scientific breakthrough has saved hundreds of thousands of lives. "I discovered over the years that I seemed to see things that other people did not see." says Kwolek. "I don't know what you attribute that to. Also I love doing chemistry. And I love making discoveries. . . . You have to be inquisitive about things. You have to have an open mind."[40]

Body Armor

Kwolek's work is a landmark in the history of troop protection. But it was not the first. From body armor to bomb detection to protective masks, science has long worked to keep soldiers safe, able-bodied, and on the battlefield. The earliest known western armor, known as the Dendra panoply, was forged in approximately 1400 BC. Made of hinged bronze plates covering the body from shoulder to past the waist, the armor was state of the art in its time. Nine hundred years later came the first chain mail, consisting of small, interlocking iron rings. Chain mail was originally made by the Celts and later adopted by the Romans, whose centurions took to wearing shirts of mail.

> **polymer**
>
> Any of numerous natural and man-made compounds of high molecular weight consisting of millions of repeated linked units.

By the thirteenth century, designers and ironworkers had added metal plates to the mail, creating a more solid protection against swords, axes, and crossbow arrows. Other protective gear followed: "They wore helmets of various shapes," says historian Yann Le Bohec, "and leather tunics . . . they carried large, flat, narrow shields."[41] Knights of medieval Europe rode onto the battlefield in full-body armor weighing between 45 and 55 pounds (20 to 25kg). The 1600s brought the evolution of heavier, thicker armor meant to protect the wearer from musket balls, but rather than armor that covered a person from head to toe, most of it was designed to shield only the most vital parts: hands, head, and torso. By 1861, when the first shots of the American Civil War were fired, metal breastplates were a luxury that few soldiers could afford. Those who could might live to see another day at Gettysburg, Chancellorsville, or Antietam.

World War II, fought between 1939 and 1945, saw the development of flak jackets, sturdy steel plates covered by thick, protective

How Body Armor Works

Bulletproof vests, such as those worn by members of the military and law enforcement, usually consist of several layers of bullet-resistant webbing (such as KEVLAR) which are then sandwiched between layers of plastic film. The combined layers are woven into an outer layer of cloth material used for making the vests.

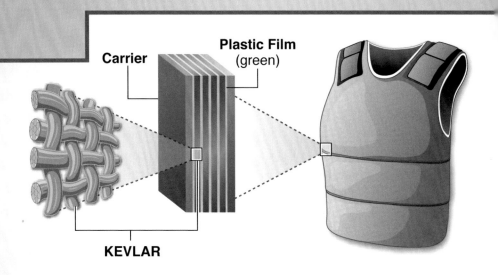

Source: How Stuff Works, "How Body Armor Works," 2011. www.howstuffworks.com.

fabric. Originally designed by the Wilkinson Sword company to stop a .45 caliber round, the jackets failed to stop higher caliber bullets and were actually far more effective in minimizing damage caused by shell blasts. During the Vietnam War many marines were required to wear the hot, 7-pound (3.2kg) jackets but they abandoned or destroyed them deeming them useless. One squad leader, according to journalist James R. Ebert, "cut the plates out of his flak jacket . . . preferring the improved mobility and lighter weight to the questionable protection the armor offered."[42] Not until the development of Kevlar did lighter and stronger protective gear become a reality.

New Materials for the Battlefield

The groundbreaking research that made Kevlar possible ended decades ago, but today at the Polytechnic Institute of New York University, another kind of body armor experimentation is being done. It may one day help engineers develop stronger, tougher, more damage-resistant armor and helmets for the military. Each day in the laboratory, materials scientist and mechanical engineer Nikhil Gupta places a variety of objects into a custom-built compression machine. The high-powered machine exerts intense pressure on everything from rabbit bones to industrial protective foam. Gupta captures every cracked, warped, and shattered object with a high-speed camera that films the objects being crushed at ten thousand frames per second.

Afterward, Gupta and colleague Paulo Coelho, a dentist and materials scientist, study the footage closely. Through their research, they have determined that an injury's severity often depends on the speed at which the bone is compressed, or squeezed. High rates of compression resulting from bomb blasts, for example, yield tiny micro-cracks rarely detected unless scanned under an electron microscope. "We were surprised to find that not only did the nature of the bone fractures change depending on the speed of compression," says Coelho, "but that bones crack in different directions based on speed."[43]

This multiple-angle approach to studying such injuries may lead to the development of protective materials—particularly the foam used in body armor and head gear—that can better withstand extreme battlefield violence. "I was already studying foams and body armor and developing new protective materials, but my approach changed when I learned about the nature and prevalence of IED [improvised explosive devices] injuries," Gupta says. "I realized it was critical to understand how the bones themselves behaved in these circumstances in order to devise the next-generation of protection."[44]

By wedding their foam research with their studies of bone compression, Gupta and Coelho hope to one day develop methods by which makers of protective gear can minimize the damage done to soldiers in the field. Thomas Killion, deputy assistant secretary of the US Army for research and technology, considers the work of these scientists and technicians critical for successful military missions. "We have to be about the current fight," says Killion. "We have to look at how do we take advantage of the great technology that is available today and bring that to bear rapidly and provide it to the Soldiers."⁴⁵

Improved Helmets

The introduction of steel helmets issued to every soldier on the Western Front during World War I marked an important innovation in protecting against head injuries on the battlefield. Those early twentieth-century helmets were not bulletproof, but they did protect members of the military from falling debris, particularly shrapnel—the tiny but deadly fragments from an exploded artillery shell. The French M15 Adrian helmet, the first modern steel helmet, weighed a light 1 pound 11 ounces (0.765 kg). By the end of World War I, more than 3 million Adrians had been manufactured and shipped to countries from Belgium to China. The American M1 helmet was designed in 1941. Its characteristic dome shape offered more protection against head injuries and, by 1945, more than 22 million were in use. The Vietnam era M1s had a less pronounced dome and were painted a lighter, olive green.

Recently improved helmets are already saving the lives of soldiers in far-flung combat zones. In the fall of 2011 the US Army issued the latest version of the enhanced combat helmet (ECH). It is made not of metal but of an ultrahigh-molecular-weight polyethylene, or plastic, that can stop rifle bullets fired at close range. Research into the ECH began in 2007, when the US Marine Corps provided a total of $8 million to four technology companies for the development and creation of the next generation of helmets. By 2010 five models of the ECH had failed to meet the standard set by the Marines: the helmet had to protect 35 percent better than an earlier model, the advanced combat helmet. Then, in the spring of 2012 a new ECH prototype made by Ceradyne, Inc., passed the required test. "We had hoped for a 35 percent improvement over the [previous model] in terms of ballistic protection and it's way

> **polyethylene**
>
> A resin (or plastic) that is being used in the development of next-generation helmets.

Scientists and engineers have developed remotely operated robots (pictured) that can disarm improvised explosive devices, or IEDs. Insurgents have used IEDs against US and other allied troops in Afghanistan and Iraq—with deadly results.

better than that,"[46] says Colonel William Cole, project manager of Soldier Protection and Individual Equipment. Years of trial and error in the laboratory resulted in the strongest helmet ever created. "The ECH is a significant technology leap forward for our Marines and Soldiers in providing increased head protection on the battlefield,"[47] says David Reed, Ceradyne vice president.

Defusing Danger

Developing the latest body armor and headgear is vital to saving lives, but contemporary researchers are also looking at ways to stop deadly attacks before they happen, particularly those caused by IEDs. Built from easily found materials such as diesel fuel, fertilizer, or parts of other, unexploded bombs, IEDs have been in use since World War II and have killed or maimed thousands. Typically placed where vehicles pass and detonated by remote control or with a wire attached to a battery, IEDs have been particularly deadly for American soldiers during recent conflicts in Iraq and Afghanistan.

To combat the use of IEDs, troops have employed bomb-defusing robots and radio jammers to block detonation frequencies. One of the latest anti-IED methods was designed by engineers Steve Todd, Chance Highs, and Juan Carlos Jakaboski. The three engineers took the idea of children's squirt guns and applied them to the serious business of saving lives. Their clear plastic Stingray, developed by Sandia National Laboratories and the company Team Technologies, literally cuts through any roadside bomb by firing a small but potent explosive charge. The charge sets off a shock wave which discharges water in a concentrated and lightning-quick stream. "The fluid blade disablement tool will be extremely useful to defeat IEDs because it penetrates the IED extremely effectively," said one Sandia project manager during the Stingray's development phase. "It's like having a much stronger and much sharper knife."[48] At a cost of only $58 per squirt gun, the Stingray is likely to be the American military's roadside bomb–killer for years to come.

A zap from the Stingray may defuse a roadside bomb before it explodes, but soldiers also find peril when insurgents attack military bases or barracks. Working with the US Army, Berry Plastics of Indiana has developed a new technology that strengthens walls and keeps them from shattering and sending fragments of concrete and metal into the air after an explosion.

Stronger than Kevlar

Kevlar revolutionized the synthetic fabric industry in the 1960s. But not until the 1980s did researchers consider its use in body armor. Since then, Kevlar has remained the standard in bulletproof vests used by law enforcement and in militaries around the world.

In 2010, though, a team of Israeli scientists from the Weizmann Institute of Science and Tel Aviv University announced that they had developed an even stronger material than Kevlar. Although it has yet to be named, the transparent material is chemically similar to the proteins that make up the plaques found in the brains of people suffering from Alzheimer's disease. The microscopic material is then reinforced with a protective layer that adds to its strength. In testing, the only way of damaging the material was with a diamond-tipped probe. (A diamond is the hardest substance known.)

The innovative technology could lead to cheaper and lighter body armor. But that day is still a ways off. It could be decades before scientists can put their findings to real use. "I think this is an amazing discovery," says scientist Kenneth Woycechowsky. "The rigidity and stiffness of these spheres is unique, and surpasses any other known organic molecule, even Kevlar."

Quoted in Eric Bland, "Toughest Body Armor Developed by Scientists," MSNBC, October 22, 2010. www.msnbc.msn.com.

The X-FLEX Blast Protection System is made from polymers, which consist of many molecules strung together to form very long chains. During installation, workers peel away a protective film liner from the wallpaper-like sheets of X-FLEX. Each sheet is backed by a sticky adhesive. After installation, the wall can be primed and painted. In case of an explosion, metal fasteners keep the sheets pinned to the wall. "The material is placed on the interior side of exterior walls and intended to protect the occupants of that particular room,"[49] says Elizabeth Curran, business development manager at Berry. Although the US military will not confirm or deny the use of X-FLEX, tests suggest that the system is stronger than the walls it is designed to protect. In one demonstration of the product, a single sheet of X-FLEX stopped a wrecking ball from penetrating a brick wall.

Safer Explosives

While much of the research being done today is designed for troop protection, some of it is also focused on improving existing weaponry by making it as safe as possible. Explosives, for example, remain a necessity in a combat zone. When used properly, they can provide an effective tool by which soldiers can take down a terrorist hideout or blow up a cache of weapons. TNT, the traditional explosive, is unpredictable. When carried and dropped, or if fired at by insurgents, TNT can sometimes set off unintentional blasts, killing or seriously injuring those nearby. In 2010 a new kind of explosive, the IMX-101, was invented and introduced by engineers at BAE Systems. It provides the same powerful blast as TNT but is more stable, less sensitive, and, therefore, safer to use and store. Researched and developed over four years, IMX-101 can also be used in artillery shells and other projectiles.

American military leaders believe that although it costs $8 per pound versus $6 for TNT, the life-saving potential of IMX-101 is more than worth it. Within ten years researchers in the United States hope that the explosive material will completely replace TNT. "IMX-101 explosive has the potential to revolutionize military ordnance," says Jerry Hammonds, vice president and general manager of BAE Systems. "This transformation will help save lives on and off the battlefield."[50]

PyroMan Project

Each year the federal government spends millions of dollars on weapons and defense research. Although some of that funding goes to private defense companies, much of it is awarded to universities doing cutting-edge work in the science of protective gear to shield soldiers from fire or harmful chemicals they may encounter in combat situations. The Textile Protection and Comfort Center at North Carolina State University (NCSU) is one institution pushing the boundaries of modern science. One typical experiment takes place in a special room heated to 1,000ºF (538ºC) by propane gas flames that lick and scorch PyroMan, a life-sized mannequin made from flame-resistant polyester resin and equipped with 122 thermal sensors.

Jadey Pareja: Testing the Next Generation of Protective Gear

When soldiers are deployed into urban combat zones they depend on scientists like Jadey Pareja. The Edgewood Chemical Biological Center (ECBC) in Aberdeen Proving Ground, Maryland, where Pareja works, is one of the many scientific units that the US Army depends on to keep its military personnel safe from dangerous materials. There Pareja and her fellow researchers work to design, engineer, and test more effective protections against harmful chemical and biological agents. Pareja spearheads one group of five scientists whose job it is to test and analyze the carbon materials used in protective mask filters.

From an early age, Pareja was fascinated by chemicals, and by the time she reached middle school the encouragement of teachers nurtured her interest into a full-time passion. Initially she trained to be a pharmacist, but after completing college, she took a job as a chemical contractor until being hired by ECBC. Pareja and her team expend much of their efforts on testing the carbon filters used on protective masks that allow soldiers to breathe in fresh air when surrounded by harmful chemicals. Other team members test the equipment itself—gloves, boots, and hazardous material suits. "Our mission within our team is to protect our protectors," Pareja says. "Everything we do within these walls is eventually going to be fielded to the Soldier. If we don't do something properly or miss one tiny step, it could ultimately affect the life of someone protecting our country."

Quoted in Dan Lafontaine, "Army Scientist Readies Soldiers' Masks for Chem-bio Hazards," *RDECOM Public Affairs*, March 5, 2012. www.army.mil.

Dressed in the latest experimental protective gear, PyroMan helps researchers determine how the clothing defends the body against extreme heat. "The copper discs in PyroMan's thermal sensors absorb the heat and tell us where he's received first-, second-, or third-degree burns," says Roger Barker, textile engineering researcher. "These data tell where we must improve the protective clothing. It's a huge benefit to anyone who wears a uniform."[51]

Tests using PyroMan and other mannequins have led directly to the creation of lightweight, highly flexible, flame- and chemical-resistant body suits, boots, and gloves. So, too, have experiments using real people. The lab is flooded with a harmless chemical—vaporized oil of wintergreen. Then, a member of Barker's team wearing a protective suit enters the lab and performs a series of tasks: climbing a ladder, opening a door, picking up a key from the floor. Inside the suit are absorbent pads which soak up any wintergreen vapor that penetrates the protective fabric. Later, the scientists analyze the amount of vapor penetration and work to keep more of it out the next time. This trial and error method is at the heart of scientific testing and ensures that over time, the team will be able to improve their designs. "All military branches have been interested in the abilities of the protective gear we design," says Barker. "They appreciate that these suits and gloves can protect military personnel from thermal hazards like those from an (improvised explosive device)."[52]

3-D Technology

Since 2008 NCSU, the University of North Carolina–Chapel Hill (UNC–CH), and Duke University have seen their defense research funding skyrocket from $4.9 million to $21.3 million. While NCSU specializes in flame- and explosive-resistant technologies, the other two universities are researching 3-D computer training models and infrared detection, which may lead to improvements in night-vision technology. At UNC–CH, researchers are concentrating on new methods of training military personnel for life in a war zone. To that end, they have developed a virtual reality 3-D imaging device that allows soldiers to see a smaller-than-life version of what they will see once deployed. Automatic cameras superimpose colorful, multidimensional graphics—buildings, trees, mountains—onto a large table filled with sand. Technicians use an electronic pen attached to a long coil to manipulate the 3-D images. Rather than simply having a mission location described to them, soldiers can get a better sense of what their surroundings will actually look like.

> **infrared**
>
> The part of the invisible spectrum that is near the red end of the visible spectrum and that comprises electromagnetic radiation of wavelengths from 800 nanometers (nm) to 1 mm.

This information could help soldiers envision themselves on the battlefield. The technology encourages them to imagine their choices and see how those choices might play out in the real world. "Right now, when

Marines train, supervisors in orange vests observe them and take notes. They're experts, but they can't see everything," says computer science researcher Greg Welch. "The military asked us to create an automated system to analyze what Marines do in a quantifiable, regular way. Actually showing an individual what he or she does wrong could save their lives."[53]

Night-Vision Goggles

Scientists from NCSU and Duke, meanwhile, are also working to improve a valuable military tool: night-vision goggles. Military personnel depend on this technology to be able to study the enemy at night. The most recent versions of the goggles are unwieldy and their clarity

A US fighter pilot adjusts his night-vision goggles. Research is leading to improvements in night-vision technology.

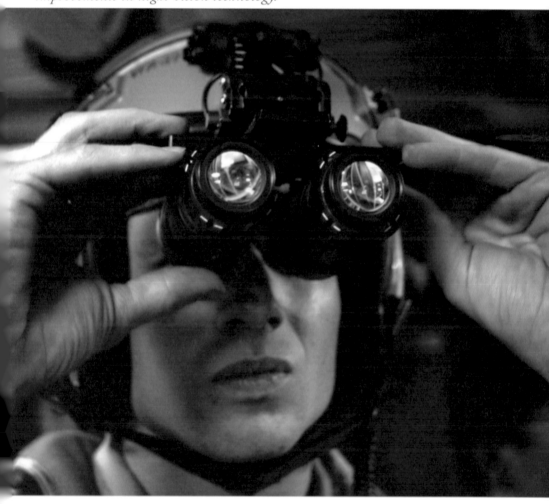

imperfect at best. But by building new goggles that contain a thermal imaging agent called vanadium oxide, the researchers hope to create a smart sensor, which, like a smartphone, is faster and more sensitive to its user. Best of all, the vanadium oxide is simply placed onto a silicon computer chip and slid into the goggles, making them lighter and easier to carry than ever before. "By putting the sensor and the computer on the same chip, we've made the device wireless," says Jay Narayan, an NCSU engineering researcher. "It's smart—it can sense, manage and respond to things quickly, especially on the battlefield."[54]

As for quality of sight, engineer Adrienne Stiff-Roberts works with semiconductors called quantum dots that measure energy types in an attempt to improve infrared detection. This technology has promise for the military, which has been forced to use low-quality infrared goggles while trailing terrorists or insurgents after dark because higher-quality models require storage in liquid nitrogen at a chilly -132°F (-91°C). Deep-freeze units are expensive and difficult to maintain in remote combat areas. Stiff-Roberts's quantum dots, on the other hand, not only provide high-quality night vision but do not need to be kept in the deep freeze.

> **quantum dot**
>
> A minuscule crystal of semiconductor material composed of various chemical compounds.

By studying, experimenting with, and designing new materials for body armor and masks, scientists are transforming the way wars will be waged for years to come. But their scientific research is also leading to breakthroughs in the development of technologies such as protective walls and safer explosives. In this way, scientists are making the world a less destructive place.

The Future of Weapons and Defense Research

The future of weapons and defense research remains limited only by the imaginations of engineers, designers, and scientists. While the methods by which war is waged have evolved over the centuries, what has not changed is the desire to create more effective ways of carrying it out. More than ever, researchers also strive to design weapons and gear that will protect those in combat. From the use of advanced robotics to launch and carry out military strikes to space-age weaponry calibrated to detect and destroy missiles launched by hostile nations or rogue states, the twenty-first century has produced the most sophisticated defense tools yet invented. Some require research into the smallest particles known to science: atoms. It is on this level—undetectable by the human eye—that scientists are exploring the next generation of weapons and crossing a new frontier in the future of war and conflict.

Robotics

Once considered the stuff of science fiction novels and movies, robotics is transforming modern weaponry in the twenty-first century. Researchers, manufacturers, and defense corporations are working to build better, stronger, and faster robots to place on the field of battle. Until now, robots have been used primarily to defuse explosives. In the future, as technology improves, they will likely take on more combat-oriented missions.

One new weapon is the MAARS (Modular Advanced Armed Robotic System), developed by QinetiQ. Looking much like a tank, MAARS is designed to be unmanned and controlled remotely. Its turret—the top part of the vehicle—rotates to enable the controller to point MAARS's powerful M240B machine gun at nearly any nearby target. It also comes equipped with machine guns, lasers, and a gripping claw used to open doors. If necessary, controllers can operate the rolling robot at night because of its night-vision capability. Its compact size—roughly as big as a riding lawn mower—makes MAARS more versatile than other military vehicles, allowing it to maneuver across almost all surfaces, from rocky hilltops to sandy beaches. Perhaps best of all, MAARS is built to stand in for flesh- and-blood soldiers, keeping them out of harm's way. "These robots can replace Soldiers in dangerous situations," says technology expert Adrian Herkenbrack. "The advantage is that these robots have no fear, and we'd rather lose one of them than a Soldier."[55]

Military operations require reconnaissance, or the ability to know where the enemy is at all times. For at least a decade, researchers have been experimenting with the concept of pocketbots—small, remotely controlled machines that can be tossed into alleyways or buildings for the purpose of detecting bad guys. In the initial tests, the bots were heavy, some weighing as much as 125 pounds (56.7kg), but like most modern technology, the latest versions are far tinier. The First Look is a robot similar in appearance to a standard tank, yet it weighs only five pounds and is the size of a hardback book. It climbs stairs and tumbles down them before zipping away and broadcasting what it sees with four built-in color cameras. Even smaller and lighter is the dumbbell-shaped Throwbot developed by researchers at Recon-Robotics. The Throwbot is designed to be tossed into a room or over a wall to detect the presence of enemy fighters. It has a titanium shell, two spoked wheels for movement, and a single black and white camera. Its range is far less that the First Look, only 300 feet (91.4m) outdoors and 100 feet (30.5m) indoors. Still, the Throwbot can provide valuable information to soldiers in the field.

> **pocketbot**
>
> A small, remotely-controlled machine that can be flown into areas for the purpose of identifying the enemy.

When land combat moves to the sea, ARIES (Acoustic Radio Interactive Exploratory Sensor) may be the underwater robotic vehicle to use. Like MAARS, it is relatively small—roughly 20 feet (6m) in length. While ARIES can be fitted with torpedoes or other weapons, it may be particularly helpful in research and underwater mine detection. A fixed-focus wide-angle video camera located in the nose of the vehicle can be used to identify underwater mines and other such objects.

A micro aerial vehicle (MAV) flies during an operational test flight. These tiny airborne vehicles are controlled remotely and might one day be used to spy on enemies or deliver messages to soldiers in the field.

Micro Aerial Vehicles

Land and sea robots will likely be an important part of any army's arsenal in the future, but researchers in Great Britain, South Korea, Israel, and the United States are also testing the next generation of MAVs, or micro aerial vehicles. These tiny airborne vehicles are controlled remotely, from across the street or across the world. Because of their size, they are

> **Robo-Mule**
>
> One of the latest developments in defense research is the robotic mule. The four-legged, mechanical LS3 may one day be widely used as a way of shouldering at least some of the burden of troops, who often have to carry 100 pounds (45kg) of gear. Still in the development stage, the mule can be programmed to follow military personnel across the rockiest terrain and may one day heed verbal commands such as "stop" and "come here." Best of all, the robot will likely be able to carry up to 400 pounds (181kg) for 20 miles (32km) without refueling. The mule's hardy ancestor, called "Bigdog," was designed by Boston Dynamics for the US military's Defense Advanced Research Projects Agency (DARPA) to withstand all weather conditions and right itself if knocked over. The much improved mule could prove a small boon for the American military when it is put into service. "If successful, this could provide real value to a squad while addressing the military's concern for unburdening troops," says army lieutenant colonel Joe Hitt. "LS3 seeks to have the responsiveness of a trained animal and the carrying capacity of a mule."
>
> Quoted in Fox News, "Robo-mule Hauls Military Gear & Follows Like a Dog," February 8, 2012. www.foxnews.com.

likely to be used to spy on enemies or to deliver messages to soldiers in the field of combat. While the technology has been under study since at least 1986, not until now has it been refined enough to make very small MAVs viable.

During a 2007 antiwar rally in Washington, DC, participants were startled and unnerved by hovering mechanical bugs that looked to some like small helicopters. No government agency or private research company took credit for the robotic creatures, but many in the crowd assumed they were being used to spy on their activities. Since then, robobug technology has continued to evolve. Researchers at the California Institute of Technology soon after developed something they called a "microbat ornithopter." The device flew on sail-like wings and was small enough to take off and land from the palm of a hand. A team of young scientists at Harvard University later devised a smaller robobug that flapped its wings at 120 beats per second. "It showed that we can manufacture the articulated, high-speed structures that you need to re-create the complex wing motions that insects produce,"[56] says team leader Robert Wood.

For other researchers, bats and birds, not insects, are the best models for MAVs. In September 2010 scientists at Boston University and the University of Maryland, among others, won a five-year, $7.5 million grant from the Office of Naval Research for a project called AIRFOILS (Animal Inspired Flight with Outer and Inner Loop Strategies). The grant is unique because it employs scientists from a variety of disciplines, including biologists who are studying how a bat or bird's physical traits, movements, and instincts could be applied to a new generation of airborne vehicles. "It allows me to do the biology I've always wanted to do," says bat biologist Tom Kunz, "but it also inspires engineers to create new aircraft."[57]

> **micro aerial vehicle**
>
> A tiny unmanned aircraft system used for surveillance purposes.

Specifically, Kunz and research partner Nickolay Hristov use 3-D thermal imaging to observe bats as they fly en masse from their caves at night to look for food. Initially, the nocturnal creatures remain clustered together as they fly, but as they pour into the sky they eventually go their separate ways. Kunz and Hristov study the bats' flight patterns, trying to understand how they move together (like cars on a crowded highway), change speeds, and suddenly change directions and disperse. If innovative researchers can pinpoint some of these patterns, habits, and skills, they may be able to apply them to the MAVs of the future.

Advanced Robotics

In March 2012 the Defense Advanced Research Projects Agency (DARPA) released a video of another robot of the future. The Cheetah robot, built by Boston Dynamics, is modeled on the fastest land animal on Earth. In the video, the Cheetah appears to be running backward, with its mechanical legs flexed in the opposite direction to that of warm-blooded animals. This unique design allows the robot to reach speeds of 18 miles per hour (29kmph); scientists hope to one day increase that to 70 miles per hour (112.7kmph). The mechanical cat's flexible spine allows it to propel itself across the ground quickly and, say experts, could be used to carry emergency supplies and aid in firefighting or other relief situations. Built into many futuristic weapons like the Cheetah will be an artificial intelligence (AI) that can calculate risks and make decisions with little human manipulation.

While the advent of more sophisticated robots equipped with AI gives many military experts hope that one day human soldiers will no longer be needed on the battlefield, others have their doubts. Without humans on hand to lead a mission or to survey the terrain or target, errors in judgment may become more common. "If the decisions are being made by a human being who has eyes on the target, whether he is sitting in a tank or miles away, the main safeguard is still there," says Tom Malinowski of Human Rights Watch. "What happens when you automate the decision? Proponents are saying that their systems are win-win, but that doesn't reassure me."[58]

Other experts acknowledge such fears but say they are overstated. "A lot of people fear artificial intelligence," says John Arquilla, executive director of the Information Operations Center at the Naval Postgraduate School. "I will stand my artificial intelligence against your human any day of the week and tell you that my A.I. will pay more attention to the rules of engagement and create fewer ethical lapses than a human force."[59]

Nano Weaponry

Fears of robotic warfare may or may not be justified, but those weapons can at least be viewed with the naked eye. The scale of nano weaponry, on the other hand, is minutely small. A centimeter is one-hundredth of a meter—quite small. A millimeter is one-thousandth of a meter—smaller still. A micrometer is one millionth of a meter—exceedingly tiny. Still, the micrometer is gigantic when compared to what is referred to by scientists as nanoscale. At this level, the primary unit of measurement is the nanometer (nm), which is one-billionth of a meter—a hundred-thousandth

> **nanoscale**
>
> An atomic world so tiny that only the world's most powerful microscopes can detect it.

A colony of bats hunts for their nighttime meal. One group of researchers is studying bats as they fly en masse from caves at night in search of food. Increased knowledge about how groups of bats move, change speed and direction, and scatter might have applications for the MAVs of the future.

the width of a human hair. Even at this infinitesimal size, the nanometer is huge in relation to atoms, the foundations of all earthly matter. The distance across an atom—its diameter—is only 0.1nm.

It is at this level that nano scientists are doing the bulk of their research, attempting to manipulate atomic particles to create never-before-seen technology. Their primary tools for such research are scanning tunneling microscopes, which use electric current to navigate through the atomic material, and atomic force microscopes that use an extra fine tip to probe the miniature matter. Also required is a computer to then scan and assemble the data. This kind of observation and experimentation has made it possible for scientists to build two exciting structures on this nanoscale: nanowires and carbon nanotubes.

The carbon nanotube is achieved by rolling a sheet of carbon atoms into a tube. Like rolling dough to make cookies, moving the roller in different directions creates different patterns. Certain patterns have been found to be particularly strong. By using the proper pattern, scientists have found that carbon nanotubes can be made hundreds of times stronger than steel yet are six times lighter. Such technological breakthroughs have applications in aviation, automobiles and electronics, and weaponry.

One potential use of this pioneering technology currently under development at DARPA is a bird-sized unmanned nano aerial vehicle (NAV). More sophisticated and far smaller than drones or MAVs, the 6.5-inch long NAV (16.5cm) will be equipped with reconnaissance sensors and be able to fly through open windows. Like a helicopter, the NAV will be able to move vertically as well as horizontally. The bionic "bird," developed by the company AeroVironment (AV) of California, is designed to look like a hummingbird and is called the Nano Scout (Nano Sensor Covert Observer in Urban Terrain.) Battery-powered and controlled remotely, the Nano Scout flaps its wings and weighs no more than two nickels. It can reach speeds up to 20 miles per hour (32kmph) or slow down to 1 mile per hour (1.6kmph) to investigate the interior of buildings.

This nanotechnology breakthrough comes only after years of hard work and persistence on the part of designers and technicians, according to AV project manager Matt Keennon: "The success of the Nano Hummingbird was highly dependent on the intense combination of creative, scientific, and artistic problem-solving skills from the many AV team members, aided by a philosophy of continuous learning, which we feel was only possible due to the unique environment here at AV."[60]

Super Suit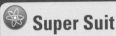

Iron Man used to be the stuff of comic books and movies, but no more. Researchers at the Raytheon Sarcos company in Salt Lake City, Utah, have recently found a way to turn science fiction into fact. The company's XOS 2 Exoskeleton is a computer-controlled suit of iron weighing 195 pounds (88.5kg). Its power to punch through walls and lift up to 200 pounds (90.7kg) comes from warmed hydraulic fluid which courses through the metal frame. "When donned, the suit seems to support itself, making its weight barely perceptible," says Raytheon-Sarcos vice president Fraser Smith. "It feels relatively lightweight, similar to a winter coat being draped across shoulders." Over the course of eight years, the US government spent $4 million to $8 million per year on the first version of the XOS, but it has yet to approve funding for the latest version. Despite the lack of government funding, researchers hope to have the XOS 2 on the market by 2015.

Quoted in *Mail Online*, "The Real Iron Man: Robotic Suit That Gives Wearer 17 Times More Strength to Be Used by Military 'Within 5 Years,'" September 30, 2010. www.dailymail.co.uk.

Nano Future

NAVs such as the Scout rely on Global Positioning Systems (GPS) to find and engage their targets, but researchers are developing vision-based systems that will enable NAVs to navigate through dense urban areas in which GPS might be unavailable. Developing these independent systems has been a challenge for scientists; the materials being used are infinitesimally tiny and therefore difficult to see and control. Many companies have looked to nature for inspiration. Specifically, insect nervous systems and brains have been carefully studied and will perhaps be applied to NAVs one day. "A lot of structures in insects are multifunctional," says Sean Humbert, an aerospace professor whose work is being used by the US Army. "Biologically, they're multitasking."[61] The ability to do many things at once could make for the most sophisticated weapons yet invented.

Futurists also envision a day in which nanotechnology may be used to manipulate the brain chemistry of its human targets. This concept of nano-poisoning could make it possible to administer a nano-filled serum that makes a prisoner tell the truth or reveal secrets, or to calm a hostile one without force. It could also be used to erase a person's memory, using minuscule micro fields that would flare in the brain.

Nano weapons could also be employed in physical combat. While it may seem the stuff of science fiction, nano needles may be fired at opponents from great distances and pin them painlessly against a wall or stop them in their tracks. Nano rifles might be fired to launch water balls that would knock an enemy down. At least one American researcher is confident that this kind of independent, self-contained nanotechnology is not that far off. "Within 10 to 15 years, autonomous Microsystems will be on the battlefield,"[62] says Neil Adams of the Draper Laboratory.

Electric Shields

Unlike nanotechnology, the future of armor worn by soldiers may use an old method of protection: electricity. Since the early 2000s British researchers have been working on a way of sending electrical current into the metal plating of tanks or body armor to repel enemy fire. Scientists at the Defence Science and Technology Laboratory (Dstl) envision a day when a tank or other military vehicle will be able to detect an incoming attack and activate an electromagnetic force field around the vehicle.

> **supercapacitor**
>
> An electrical component capable of holding hundreds of times more electrical charge than a standard capacitor, which consists of an electric circuit element that temporarily stores charge.

The electric shield would last for only a fraction of a second but if timed correctly this would be enough to minimize or completely repel rocket-propelled grenades (RPGs), which explode on impact. The force field would get its energy from what scientists call a supercapacitor, a device that would store the electric energy in preparation for use. "The supercapacitor material can be charged up and then discharged in one powerful event," says Dstl scientist Bryn James. "You would think this would require huge amounts of energy, but we have found it can be done with surprisingly small amounts of electrical power."[63]

One advantage of this kind of defense system is that it could have the effect of making future tanks lighter and more versatile. Today's tanks are heavy, bulky, and loaded with enough armor to minimize the damage

The tiny drone aircraft known as the Nano Hummingbird demonstrates its abilities during a press briefing. It weighs less than two nickels and can reach speeds up to 20 miles per hour.

done by RPGs, which can penetrate a foot of steel. Electric armor could therefore revolutionize the design and construction of armored vehicles.

Space Weapons

Throughout the history of aviation, airplanes and spaceships have remained separate and distinct airborne vehicles. But the future of weapons and defense research will likely blur that line. In May 2010 scientists with the US Air Force launched an experimental vehicle called an X-51A Waverider into the sky. Propelled by a rocket booster and an air-breathing scramjet, which takes oxygen from the atmosphere rather than from onboard tanks, the Waverider reached the speed of Mach 5 and became the longest hypersonic flight ever powered by this kind of engine.

For researchers in the field this was a technological breakthrough that may one day pave the way for space planes that can travel 600 miles (965.6km) in only ten minutes. Experts imagine that these kinds of space planes may one day be used in wars taking place high in the stratosphere. One such craft, the X-37B orbital test vehicle, was launched in March 2011. The unmanned X-37B, which resembles the space shuttle, was supposed to remain in orbit for nine months. More than a year later, it continued circling the earth at 17,000 miles per hour (27,359kmph) as researchers tested its secret capabilities. Solar-powered and measuring 29 feet (8.8m) in length, the X-37B is clearly meant to be the next generation of space plane. "We initially planned for a nine-month mission," says program director Lieutenant Colonel Tom McIntyre. "Keeping the X-37 in orbit will provide us with additional experimentation opportunities and allow us to extract the maximum value out of the mission."[64]

The Quest Continues

Whatever the fate of the X-37B may be, visionary science fiction authors have long predicted that the next frontier of war will be outer space. *Earthlight*, a 1955 novel by Arthur C. Clarke, introduced the concept of a device that uses electromagnets to shoot high-speed streams of molten metal at enemy targets. Today, DARPA researchers are developing such a weapon called MAHEM (Magneto Hydrodynamic Explosive Munition). While no official timetable for the space weapon has been made public, the introduction of MAHEM may be the first salvo in a future space war.

When—or even whether—such technology will be usable is difficult to say. "I would say that within the next 15 to 30 years . . . but probably 15 to 20 years, you could start to see this technology being expanded to the point where you could get aircraft into outer space,"[65] says Joseph Vogel, hypersonics director at Boeing.

Whatever the future may hold for weapons and defense research, the quest to build more effective combat tools will continue. In a world in which conventional warfare gives way to robotics, nanotechnology, and space weaponry, technicians and scientists will be more essential than ever before.

Source Notes

Introduction: Confronting Threats in the Twenty-First Century

1. Quoted in Michael Murray, "The Navy SEAL Team 6 Weapons and Gadgets That Brought Down Osama bin Laden," ABC News, May 3, 2011. http://abcnews.go.com.
2. Quoted in Murray, "The Navy SEAL Team 6 Weapons and Gadgets That Brought Down Osama bin Laden."

Chapter One: What Is Weapons and Defense Research?

3. Quoted in Donald Simanek, *Science Quotes*, Lock Haven University, 2001. www.lhup.edu.
4. Quoted in Central Intelligence Agency, "The Bureau of Military Information," January 3, 2012. www.cia.gov.
5. Quoted in Spencer Tucker, *Tanks: An Illustrated History of Their Impact*. Santa Barbara: ABC-CLIO, 2004, p. 15.
6. Quoted in Max Arthur, *The Faces of World War I*. London: Cassell Illustrated, 2007, p. 122.
7. Quoted in Geoffrey C. Ward, *The War: An Intimate History, 1941–1945*. New York: Knopf, 2007, p. 84.
8. Quoted in Ward, *The War: An Intimate History, 1941-1945*, p. 116.
9. Chris Bishop, *The Encyclopedia of Weapons of World War II*. New York: Metro, 1998, p. 7.
10. Quoted in Steven Lehrer, *Wannsee House and the Holocaust*. Jefferson, NC: McFarland and Company, 2000, p. 11.
11. Quoted in Lehrer, *Wannsee House and the Holocaust*, p. 10.
12. Quoted in G. Wayne Miller, *The Xeno Chronicles*. New York: Public Affairs, p. 69.

Chapter Two: Drones

13. Quoted in Thomas Harding, "Col. Gaddafi Killed: Convoy Bombed by Drone Flown by Pilot in Las Vegas," *Telegraph* (UK), October 20, 2011. www.telegraph.co.uk.

14. National Aeronautics and Space Administration, "Aeronautics and Astronautics Chronology, 1915–1919," 2005. www.hq.nasa.gov.

15. Taylor Downing, *Churchill's War Lab*, online ed. New York: Penguin, 2011.

16. Quoted in Noah Shachtman, "Computer Virus Hits U.S. Drone Fleet," CNN, October 10, 2011. www.cnn.com.

17. Quoted in Nick Fielding, "US Draws Up Plans for Nuclear Drones," *Guardian* (Manchester, UK), April 2, 2012. www.guardian.co.uk.

18. Quoted in Keith Rogers, "Dad Bitter over Son's Friendly Fire Drone Death," *Las Vegas Review-Journal*, April 3, 2012. www.military.com.

19. Quoted in Declan Walsh, Eric Schmitt, and Ihsanullah Tipu Mehsud, "Drones at Issue as U.S. Rebuilds Ties to Pakistan, *New York Times*, March 18, 2012. www.nytimes.com.

20. Quoted in "China Building an Army of Unmanned Military Drones to Rival the U.S.," *Mail Online*, July 5, 2011. www.dailymail.co.uk.

Chapter Three: Cyberwarfare

21. Quoted in Paul Marks, "Why the Stuxnet Worm Is Like Nothing We've Seen Before," *New Scientist*, January 18, 2011. www.newscientist.com.

22. Quoted in Holger Stark, "Stuxnet Virus Opens New Era of Cyber War," *Spiegel*, August 8, 2011. www.spiegel.de.

23. Quoted in CBS News, "Cyber War: Sabotaging the System," November 8, 2009. www.cbsnews.com.

24. Richard A. Clarke, *Cyber War*. New York: HarperCollins, 2012, p. 6.

25. Quoted in NPR, "The 'Worm' That Could Bring Down the Internet," September 27, 2011. www.npr.org.

26. Quoted in Frank Jack Daniel, "Fake Memo but Real Code? India-U.S. Hacking Mystery Deepens," Reuters, January 11, 2012. www.reuters.com.

27. Quoted in Kathleen Hickey, "How International Cyber Crime Threatens National Security," *Government Computer News*, July 27, 2011. http://gcn.com.

28. Quoted in Sarah Dingle, "Millions of South Koreans Victims of Social Network Hack," *The World Today*, ABC News, July 29, 2011. www.abc.net.au.

29. Quoted in Jessica Silver-Greenberg and Nelson D. Schwartz, "MasterCard and Visa Look into Possible Data Attack," *New York Times*, March 30, 2012. www.nytimes.com.

30. Quoted in CBS News, "Cyber War: Sabotaging the System," November 8, 2009. www.cbsnews.com.

31. Quoted in Siobhan Gorman and Julian E. Barnes, "Cyber Combat: Act of War," *Wall Street Journal*, May 30, 2011. http://online.wsj.com.

32. Quoted in Larry Greenemeier, "The Fog of Cyberwar: What Are the Rules of Engagement?" *Scientific American*, June 13, 2011. www.scientificamerican.com.

33. Quoted in Devlin Barrett, "U.S. Outgunned in Hacker War," *Wall Street Journal*, March 28, 2012. http://online.wsj.com.

34. Quoted in Greenemeier, "The Fog of Cyberwar."

35. Quoted in Barrett, "U.S. Outgunned in Hacker War."

36. Quoted in Nick Hopkins, "US and China Engage in Cyber War Games," *Guardian* (Manchester, UK), April 16, 2012. www.guardian.co.uk.

37. Quoted in Hopkins, "US and China Engage in Cyber War Games."

38. Quoted in Simon Tisdall, "Cyber-warfare 'Is Growing Threat,'" *Guardian* (Manchester, UK), February 3, 2010. www.guardian.co.uk.

Chapter Four: Protecting Troops

39. Quoted in David Brown, "Insight: Stephanie Kwolek," interview, Lemel-MIT Program, November 2000. web.mit.edu.

40. Quoted in Brown, "Insight: Stephanie Kwolek."

41. Yann Le Bohec, *Imperial Roman Army*. New York: Routledge, 2000, p. 123

42. Quoted in James R. Ebert, *A Life in a Year: The American Infantryman in Vietnam*. New York: Presidio, 2004, p. 215.

43. Quoted in Hallie Deaktor Kapner, "Bone-Crushing Experiments Yield Better Protective Gear," Live Science, September 24, 2010. www.livescience.com.

44. Quoted in Kapner, "Bone-Crushing Experiments Yield Better Protective Gear."

45. Quoted in USArmy, "Army's Research Budget to Focus on Protecting Soldiers," *Defence Talk*, March 2, 2010. www.defencetalk.com.

46. Quoted in Lance M. Bacon, "Bullet-Stopping Helmet, New Boots to War Zone," *Army Times*, February 26, 2011. www.armytimes.com.

47. Quoted in Yahoo! Finance, "Ceradyne, Inc. Receives Second Enhanced Combat Helmet (ECH) Award," May 15, 2012. http://finance.yahoo.com.

48. Quoted in Spencer Ackerman, "New Way to Stop Roadside Bombs: Super-Soak 'Em," *Wired*, September 13, 2010. www.wired.com.

49. Quoted in Larry Greenemeier, "Sticky Savior: U.S. Army Readies a New Blast-Protection Adhesive for Deployment," *Scientific American*, December 18, 2008. www.scientificamerican.com.

50. Quoted in BAE Systems, "IMX-101 Explosive Approved to Replace TNT in US Army Artillery," *Defence Talk*, July 28, 2010. www.defencetalk.com.

51. Quoted in Whitney L.J. Howell, "N.C. Scientists Helping Soldiers," *News and Observer* (Raleigh, NC), August 30, 2010. www.newsobserver.com.

52. Quoted in Howell, "N.C. Scientists Helping Soldiers."

53. Quoted in Howell, "N.C. Scientists Helping Soldiers."

54. Quoted in Howell, "N.C. Scientists Helping Soldiers."

Chapter Five: The Future of Weapons and Defense Research

55. Quoted in Jean Dubiel, "Robots Can Stand In for Soldiers During Risky Missions," *U.S. Army*, August 11, 2008. www.army.mil.

56. Quoted in Rick Weiss, "Dragonfly or Insect Spy? Scientists at Work on Robobugs," *Washington Post*, October 9, 2007. www.washingtonpost.com.

57. Quoted in Rebecca Boyle, "Bat Research Inspires Disciplines Far Beyond Biology," *Popular Science*, November 2, 2010. www.popsci.com.

58. Quoted in John Markoff, "War Machines: Recruiting Robots for Combat," *New York Times*, November 27, 2010. www.nytimes.com.

59. Quoted in Markoff, "War Machines: Recruiting Robots for Combat."

60. Quoted in AeroVironment, "AeroVironment Develops World's First Fully Operational Life-Size Hummingbird-Like Unmanned Aircraft for DARPA," February 17, 2011. www.avinc.com.

61. Quoted in Ned Smith, "Military Plans Hummingbird-Sized Spies," MSNBC, July 2, 2010. www.msnbc.msn.com.

62. Quoted in Smith, "Military Plans Hummingbird-Sized Spies."

63. Quoted in Richard Gray, "Star Trek–Style Force-Field Armour Being Developed by Military Scientists," *Telegraph* (UK), March 20, 2010. www.telegraph.co.uk.

64. Quoted in Ted Thornhill, "Revealed: How America's Secret Space Plane Has Been in Orbit for Over a Year—and No One Knows What It's Doing," *Mail Online*, March 8, 2012. www.dailymail.co.uk.

65. Quoted in Jeremy Hsu, "Air Force Sees Hypersonic Weapons and Spaceships in the Future," Space.com, June 17, 2010. www.space.com.

Facts About Weapons and Defense

Ground Weapons

- Types of artillery include antiaircraft and antitank guns (which fire at high muzzle velocity through long barrels at flat trajectories) and howitzers (with shorter barrels, lower velocities, and parabolic trajectories).

- Basic tank design has not changed significantly since World War II, although communication equipment in tanks, armor guns and targeting, and crew comfort have improved.

- The M1 Abrams' tank engine pumps out over fifteen hundred horsepower, more than the world's fastest car.

Air Weapons

- According to the *Washington Times*, by 2020 there may be as many as thirty thousand drones flying over the United States.

- In the United Kingdom, manufacturers have suggested painting drones bright colors to make them seem friendlier and less warlike.

- The US Air Force has sixty-five thousand to seventy-five thousand people working to process drone data and footage.

- While most military experts use the acronym UAVs (Unmanned Aerial Vehicles) to describe drones, the US Air Force prefers the term RPAs (Remotely Piloted Air Systems).

Cyberweapons

- Between the Defense Department and Homeland Security, the United States will spend $10.5 billion for cybersecurity by 2015.

- The Pentagon's new Cyber Command will have a staff of ten thousand.

- According to the *New York Times*, the United States and Israel designed the "Stuxnet" virus that has infected some thirty thousand computers in Iran and set back Tehran's nuclear program.

- To build a botnet, hackers send out a program disguised as a link or hidden in an e-mail attachment; this infects the host computer and communicates with a command machine.

Missiles

- The former Soviet Union completed the first operative Intercontinental Ballistic Missiles (ICBMs) in 1958; in a race to compete, the United States built its own ICBMs and gained missile superiority by 1962.

- The intermediate-range ballistic missile (IRBM) can reach targets up to fifteen hundred nautical miles away; the intercontinental ballistic missile (ICBM) has a range of many thousands of miles.

- The US Navy has a stockpile of thirty-five hundred Tomahawk cruise missiles, with a combined worth of $2.6 billion.

Nuclear Weapons

- Nuclear weapons come in two types: nuclear fission bombs (atomic bombs) or nuclear fusion bombs (hydrogen bombs).

- In today's dollars, the first nuclear bomb built cost $20 billion.

- Between 1945 and 1992, the United States detonated 1,125 nuclear devices in tests.

- From 1951 to 2012, the United States has built 67,500 nuclear missiles.

- The nuclear arsenal of the United Kingdom has enough power to destroy over 80 percent of the 195 capital cities of the world.

Related Organizations

Boston Dynamics
78 Fourth Ave.
Waltham, MA 02451-7507
phone: (781) 663-0586
website: www.bostondynamics.com

Boston Dynamics, an engineering company that specializes in building robots and software for human simulation, began at the Massachusetts Institute of Technology in 1992 and now creates robots that include BigDog, a quadruped robot for travel on rough-terrain, and RISE, a robot that climbs vertical surfaces.

DARPA (Defense Advanced Research Projects Agency)
675 N. Randolph St.
Arlington, VA 22203-2114
phone: (703) 526-6630
website: www.darpa.mil

Established in 1958, DARPA, according to its website, commissions and builds weapons and defense technology with the goal of maintaining technological superiority over US adversaries.

Human Rights Watch
350 Fifth Ave., 34th Floor
New York, NY 10118-3299
phone: (212) 290-4700
website: www.hrw.org

As one of the world's leading organizations dedicated to defending and protecting human rights, Human Rights Watch focuses international attention on the violations of such rights that occur during war.

Ploughshares Fund

1808 Wedemeyer St., Suite 200
The Presidio of San Francisco
San Francisco, CA 94129
phone: (415) 668-2244
website: www.ploughshares.org

Founded in 1981, Ploughshares Fund supports a community of experts, advocates, and analysts working to discourage the use of, and one day eradicate, nuclear weapons.

Polytechnic Institute of New York University

6 MetroTech Center
Brooklyn, NY 11201
phone: (718) 260-3050
website: www.poly.edu

Polytechnic Institute of New York University is a school of engineering, applied sciences, technology, and research. Many of the institute's students and graduates engage in defense and weapons research for the US government.

QinetiQ

North America Headquarters
7918 Jones Branch Dr., Suite 350
McLean, VA 22102
phone: (703) 752-9595
website: www.qinetiq.com

UK-based QinetiQ develops new military and space technologies, employing thirteen thousand people worldwide. With its expertise in robotics and other technologies for military use, the company and its scientists are building the latest weapons and defense systems.

Sandia National Laboratories

1515 Eubank Blvd. SE
Albuquerque, NM 87123
phone: (505) 845-0011
website: www.sandia.gov

Sandia National Laboratories, a subsidiary of the Lockheed Martin Corporation, specializes in science-based technologies related to the safe stockpiling of nuclear weapons, as well as the reduction of weapons of mass destruction and the overall defense of the United States

US Cyber Command

phone: (800) 225-5779

website: www.arcyber.army.mil

USCYBERCOM plans and coordinates cyberdefense and cyberattacks for the Department of Defense. Most of the agency's work is carried out in secret.

For Further Research

Books

Jason Andress and Steve Winterfeld, *Cyber Warfare: Techniques, Tactics and Tools for Security Practitioners*. Waltham: Syngress, 2011.

Richard A. Clarke, *Cyber War: The Next Threat to National Security and What to Do About It*. New York: Ecco, 2012.

Roger Ford et al. *Weapon: A Visual History of Arms and Armor*. New York: DK, 2010.

Matt J. Martin and Charles W. Sasser, *Predator: The Remote-Control Air War Over Iraq and Afghanistan: A Pilot's Story*. Minneapolis: Zenith, 2010.

Edward M. Spiers, *A History of Biological and Chemical Weapons*. London: Reaktion, 2010.

Benjamin Sutherland, *Modern Warfare, Intelligence and Deterrence: The Technologies That Are Transforming Them*. Hoboken, NJ: Wiley, 2012.

Websites

Future Military Weapon Technology (www.futurefirepower.com). For a peek into the future of warfare, check out this straightforward and informative website. Learn about the sleek F-22 Raptor, the first fighter jet designed completely by computer, or read about the latest weapons innovation in Russia and China.

Military Channel (http://military.discovery.com). This companion website to the cable television channel is a solid starting place for newcomers to military matters. American wars and particular battles are profiled, and weapon aficionados can watch videos detailing the latest advances in tanks, machine guns, and aircraft.

Military Factory (www.militaryfactory.com). This site features dozens of articles, illustrations, and photographs, including a detailed look at the aircraft used by the Japanese to attack Pearl Harbor in 1941 and the weaponry employed by the US Navy SEALs.

On Being a Scientist: A Guide to Responsible Conduct in Research (www.nap.edu/openbook.php?record_id=12192&page=R1). This is a free, downloadable book from the National Academy of Sciences Committee on Science, Engineering, and Public Policy. The 2009 edition provides a clear explanation of the responsible conduct of scientific research. Chapters on treatment of data, mistakes and negligence, the scientist's role in society, and other topics offer invaluable insight for student researchers.

US Army Research Laboratory (www.arl.army.mil/www/default.cfm). This website is the only way to visit a US Army Research Laboratory (ARL) facility without a security clearance. The website has videos of the lab's latest innovations and information on career opportunities in the field of weapons and defense research.

Index.

Note: Boldface page numbers indicate illustrations.

Adams, Neil, 73–74
aerial bombardment, 20–21, **21**
aerial torpedoes, 28–29
Aero Vironment (AV) of California, 72
aircraft
   antiaircraft guns, 18
   bombardment by, 20–21, **21**
   history of development of, 18–21
   MAVs, 67–69, **67**
   NAVs, 72–73, 75
   in space, 75–76
   stealth, 31, 33
   World War I, 20
   See also drones
AIRFOILS (Animal Inspired Flight with Outer and Inner Loop Strategies), 69
AK-47 assault rifles, 22
Akkad, 13–14
American Civil Liberties Union (ACLU), 37
American Civil War, 14, 28, 53
American Revolution, 14
Anderson, Kenneth, 39
Anonymous, 45
antiaircraft guns, 18
antiradar paint, 31
AQM-34 Firebees, 31–35
Arab Spring, 27
ARIES (Acoustic Radio Interactive Exploratory Sensor), 66
Arquilla, John, 69
artificial intelligence (A.I.), 69–70
artillery, 17–18, **17**, 76
ASN Technology Group, 39
atomic bombs, 19
Aviatik airplanes, 20

BAE Systems, 60
ballistics, defined, 12
balloons, 28
banks, 44, 48
Barker, Roger, 61, 62
Bees, 29–30
Berry Plastics Corporation, 35, 59
Bigdog, 68
Big Willie, 15
bin Laden, Osama, 10–11
Bishop, Chris, 21
body armor
   future, 55–56, 59, 73, 74
   helmets, 53, 56, 58
   history, 52–54, 56
   operation of, **54**
   protective masks, 61
   thermal and chemical protection, 60–62
Bohec, Yann Le, 53
Bosch, Carl, 22–23
Boston Dynamics, 68, 69–70
Boston University, 69
bots/botnets/botmasters/bot herders, 42–43, 44, 66
Bourgeoys, Marin le, 14
Bowden, Mark, 43, 48
breastplates, 53
breech loading of cannons, 17
Browning .30-Caliber Air-Cooled Machine Gun, 16
B-17 bombers, 30
bulletproof vests, 52–53, 59
Bush, George W., 32

cannons, 17
carbon filters, 61
carbon nanotubes, 72
Central Intelligence Agency (CIA), 32, 33
Cerdyne, Inc., 56, 58
chain mail, 53
Cheetah bots, 69–70
chemical weapons

nano-poisons, 73–74
napalm, 24, **25**
poison gas, 23
protective body armor, 60–62
protective masks, 61
China, 38–39, 49, 50
China Aerospace Science and Industry Corporation, 39
Chinese Aerospace Science and Technology Corporation, 39
chlorine gas, 23
Churchill, Winston, 15
Clarke, Arthur C., 76
Clarke, Richard A., 41
Coelho, Paulo, 55
Cole, Chris, 35
Cole, William, 56, 58
Colt M-16, 16
computer-assisted weapons, 34, 64, 73
computer networks, infiltration of, 41–43, 46
Conficker, 48
credit card information, 44, 45–46
crime, 44–46, 48
Curran, Elizabeth, 59
cyberattacks, 49
cyberespionage, 41–43
cyberwarfare
    attack guidelines, 41
    bots and, 42–43
    Conficker and, 48
    cyberattack vs., 49
    cyberespionage, 41–43
    defined, 41
    difficulty determining perpetrators of, 49
    ethics of, 32, **36**, 37–38, 68
    infrastructure sabotage as, 47–48
    Stuxnet and, 40

debit card information, 45–46
Defence Science and Technology Laboratory (Dstl), 74
Defense Advanced Research Projects Agency (DARPA), 62, 68, 69, 76
Deimling, Berthold von, 23
Dendra panoply, 53
DH.82B Queen Bee, 29–30

distance warfare, 35–38
dolphins, 66
Downing, Taylor, 30
drones
    civilians casualties, **36**, 37–38
    combat, 31
    domestic use of, 37, 68
    ethics and legality of use in warfare, 32
    foreign, 32–33, 38–39
    friendly fire incidents and, 35–36
    future of, **38**, 39
    history of, 28–30
    in Libyan Civil War, 27
    MAV, 67–69, 67
    nuclear-powered, 35
    remotely piloted, 31, 33–34
    surveillance, 31, 33
    viruses targeting, 34
    without human controllers, 38
Duke University, 62–63
Dunlap, Charles, 48

early history of weapons, 14
Earthlight (Clarke), 76
Ebert, James R., 54
Edgewood Chemical Biological Center (ECBC), 61
Edgeworth, Richard, 14
Einstein, Albert, 19
electric shields, 74–75
electromagnetic artillery, 76
enchanced combat helmet (ECH), 56, 58
encryption, defined, 34
encryption failures, 34
endpoint machines, 44
Enfield rifles, 16
ethics of cyberwarfare, 32, **36**, 37–38, 68
explosives
    molten metal, 76
    robots that disarm, **57**
    safety of, 60

Fairey Queen, 29
federal law enforcement domestic drone use, 37
Fieseler Fi-103, 30
Fieser, Louis, 24

Firebee, 31–33
firewalls, 41–42
Fitzgerald, Patrick, 40
"flak jackets," 53–54
"Flying Fortresses," 30
foam protective gear, 55–56
Fokker, Anthony, 20
friendly fire incidents, 31–32
fuselage, defined, 29

Gadhafi, Muammar, 27
Garand, John, 16
Gárate, Daniel, 37
General Atomics Aeronautical Systems, 33
George, David Lloyd, 15
Global Hawks, 34
Gloster Metero, 21
Gosslau, Fritz, 30
Grant, Ulysses S., 14
Great Britain
  electric shield research, 74
  MAV testing, 67
  and mortars, 18
  and unmanned aircraft, 29–30
  use of tanks by, 14–15
  during World War II, 20–21, **21**, 30
Gupta, Nikhil, 55
gyroscopes, 28, 29

Haber, Fritz, 22–24
Haber, Ludwig, 23–24
hacking
  of business and institutional information, 44
  delayed detection of, 46
  of national infrastructure, 41, 46–48, **47**
  of personal information, 44–46
  ring, government infiltration of, 45
Hammonds, Jerry, 60
Hankey, Maurice, 15
Hawks, 34
He-178, 21
Heartland Payment Systems, 46
helicopters, 10–11, **11**
helmets, 53, 56, 58
Henry, Shawn, 49–50

Herkenbrack, Adrian, 65–66
Hewitt, Peter Cooper, 28
Highs, Chance, 58
Hiroshima, Japan, 19
Hitt, Joe, 68
howitzers, **17**, 18
Hristov, Nickolay, 69
human error, 73
Humbert, Sean, 73
hydrogen balloons, unmanned, 28
Hypponen, Mikko, 45

identity theft, 44–46, 48
improvised explosive devices (IEDs), robots to disarm, **57**
IMX-101, 60
India, 39, 43
infrared, defined, 62
infrastructure
  breaching, 41, 46–48, **47**
  defined, 41
Internet, 43, 48
interrupter gear, 20
Iran, 40
Iron Man, 73
Israel
  body armor development in, 59
  and drones, 32–33
  testing MAVs, 67
  Stuxnet and, 40

Jakaboski, Juan Carlos, 58
James, Bryn, 74
Japan, 19
jet fighters, 21

Kahaner, Larry, 22
Keennon, Matt, 72
Kellerman, Tom, 46
Kettering, Charles, 29
Kettering Bugs, 29
Kevlar, 52–53, 59
Khan, Cahudhry Nisar Ali, 37–38
Killion, Thomas, 55–56
Kalashnikov, Mikhail, 22
Korean War, 16
Kunz, Tom, 69
Kwolek, Stephanie, 52, 53

Larynx, 29
legality of drone use, 32
Lewis, Jim, 50
Liang Guanglie, 50
Libyan Civil War, 27
Little Willie, 15
local law enforcement use of drones, 37
Los Alamos, New Mexico, 19
LS3, 68
Lusser, Robert, 30
Lynn, William, 51

M1A1 antiaircraft guns, 18
MAARS (Modular Advanced Armed Robotic System), 65–66
MacGibbon, Alastair, 44–45
machine guns
    electromagnetic, 76
    placed in aircraft, 20
    World War II Browning, 16
Machowicz, Richard, 10, 12
M15 Adrian helmet, 56
MAHEM (Magneto Hydrodynamic Explosive Munitions), 76
Manhattan Project, 19
Mark I, 15–16
Mastercard, 44–46
McConnell, Mike, 46
McIntyre, Tom, 76
mechanical bugs, 67–69, **67**
Meridor, Dan, 40
Mesopotamia, 13–14
Messerschmitt Bf-109Es, 20
M-1 Garand Carbine, 16
M1 helmet, 56
micro aerial vehicles (MAVs), 67–69, **67**
mission, relationship to target and weapons of, 10
molten metal explosives, 76
Monsegur, Hector Xavier, 45
Moonlight Maze, 43
mortars, developed, 18
Mother (tank), 15
"motor-war car," 14
mountain guns, 18
mules, 68
muskets, 14

Nagasaki, Japan, 19
Nano Hummingbird, 72, **75**
nano-poisons, 73–74
nanoscale, defined, 70
Nano Scout (Nano Sensor Covert Observer in Urban Terrain), 72–73
nano weapons, 70, 72
napalm, 24, **25**
Narayan, Jay, 64
national security, threats to
    ability to meet, 49–50
    cyberespionage, 41–43
    encryption failure, 34
    system sabotage, 41, 46–48, **47**
NAV, 72–73, **75**
Navy SEALs, 10–11
Nicol, David M., 49
night-vision goggles, 62, 63–64, **63**
nitric acid, 23
North Carolina State University (NCSU), 60–63
Northrup Grumman, 34, 35
nuclear-powered drones, 35

Obama, Barack, 32, 41
Oppenheimer, Robert, 19
OTO Melara howitzers, 18

Pakistan and drones, **36**, 37–38, 39
Pareja, Jadey, 61
phosphorus, defined, 24
PlayStation users, 44
pocketbots, 66
poison gas, 23
polyethylene, defined, 56
polymer, defined, 53
power grid breaches, 41, 46–48, **47**
Predator drones, **31**, 33–34, 36
privacy issues, 37, 68
prototypes, defined, 15
PyroMan Project, 60–62

Q-2C Firebee, 31
QinetiQ, 65–66
quantum dots, 64
Queen Bee, 29–30

radio-controlled aircraft, 29

Rast, Benjamin David, 36
Rast, Robert, 36
Raytheon Sarcos, 73
Real Sabu, The, 45
Reapers, 34, 35
reconnaissance
   drones, 31, 33
   MAV, 67–69, **67**
   pocketbots, 66
ReconRobotIcs, 66
Reed, David, 58
remote control weapons: development of aerial torpedoes, 28–29
   See also drones; robotics
Remotely Piloted Aircraft (RPAs), **31**, 33–34
rifles
   barrel technology of, 17
   nano, 73–74
   Vietnam War, 16
   World War II and Korean War, 16, 22
robots
   IED disarming, **57**
   micro aerial, 67–69, **67**
   for reconnaissance, 66
   as supply carriers, 68, 69–70
   tank-like, 65–66
   underwater, 66
Roosevelt, Franklin D., 19, 20
rootkits, 42, 43
root servers, defined, 46
Royal Air Force (RAF), 20–21, **21**
RQ-1 Predator, **31**, 33–34
RQ-9 Reaper, 34
Russia
   drone development in, 39
   as origin of cyberattacks, 43, 49
   World War II, 22

sabotage, 46–48, **47**
Samuel, Cherian, 43
Sandia National Laboratories, 35, 58
Sargon the Great, 14
Saydjari, O. Sami, 49
science, defining, 13
Scout, 33
scramjets, air-breathing, 75
"screw guns," 18

SEALs, 10–11
Simms, Frederick, 14
SK Communications, 44
smart sensors, 64
Smith, Fraser, 73
Smith, Jeremy D., 36
Sony, 44
South Korea, 67
Soviet Union, 22
   See also Russia
space weapons, 75–76
Sperry, Elmer, 28, 29
Sperry, Lawrence, 28
squirt guns, 35
state law enforcement use of drones, 37
Stiff-Roberts, Adrienne, 64
Stingray squirt guns, 35
Stuxnet, 40
Sudayev, Alexey Ivanovich, 22
supercapacitors, 74
surveillance
   drones, 31, 33
   MAVs, 67–69, **67**
   pocketbots, 66
Swinton, Ernest, 15

tanks
   early, 14
   electric shields for, 74–75
   World War I, 15
target, relationship to mission and weapons of, 10
Team Technologies, 35
Tel Aviv University, 59
terrorism threat
   drones used to defend against, 32, **36**, 37–38
   and mission to get bin Laden, 10–11, **11**
Textile Protection and Comfort Center, 60–62
3-D imaging devices, 62–63
Throwbots, 66
TNT, 60
Todd, Steve, 58
Tritton, William, 15
troop protection. *See* body armor

"ultra-persistence technologies," 35

United States
   cyberdefense spending by, 50–51
   as origin of cyberattacks, 49
   war games with China, 50
University of Maryland, 69
University of North Carolina–Chapel Hill, 62–63
unmanned aerial vehicles (UAVs), 28–30
   See also drones
US Cyber Command (USCYBERCOM), 41, **42**

vanadium oxide, 64
van Niel, Cornelius Bernardus, 13
Vergeltungswaffe (revenge weapon)-1 (V-1), 30
Vietnam War
   aircraft, 32
   chemical warfare, 24, **25**
   protective gear, 54, 56
   rifles, 16
viruses (computer)
   bots and, 42
   Conficker, 48
   guidelines for planting, 41
   Stuxnet, 40
   targeting drones, 34
Visa, 44–46
Vogel, Joseph, 76
Voisin V89 airplanes, 20

warfare, distant, 28–29, 35–38
   See also drones; robotics
Waveriders, 75–76

weapons, relationship to mission and target of, 10
Weizmann Institute of Science, 59
Welch, Greg, 62–63
Westmann, Stefan, 18
Wilkinson Sword Company, 54
Wilkinson, T.S., 28, 29
Wilson, Walter, 15
Wood, Robert, 69
World War I
   aircraft, 20
   artillery, **17**, 18
   chemical weapons, 23
   helmets, 56
   tanks, 15–16
World War II
   aircraft, 20–21, **21**
   artillery, 16, 18
   atomic bombs, 19
   chemical weapons, 24
   "flak jackets," 53–54
   rockets, 30
   V-1, attacks by and against, 30
worms (computer), 40, 48
Woycechowsky, Kenneth, 59
Wright brothers, 18

X-51A Waveriders, 75–76
X-FLEX Blast Protection System, 59–60
XOS 2 Exoskeletons, 73
X-37B orbital vehicles, 76

Yom Kippur War, 33

Picture Credits

Cover: iStockphoto.com

Maury Aaseng: 38, 54

AP Images: 25, 31, 42

© Bettmann/Corbis: 17

© B.K. Bangash/AP/Corbis: 36

© HO/Reuters/Corbis: 11

© Hulton-Deutsch Collection/Corbis: 21

Landov Media: 57, 63

© Reed Saxon/AP/Corbis: 75

© Kenneth G. Takada/Zuma Press/Corbis: 67

Thinkstock: 8, 9, 47, 71

About the Author

David Robson's many books for young people include *Colonial America, The Decade of the 2000s,* and *Encounters with Vampires*. He is also an award-winning playwright whose work for the stage has been performed across the country and abroad. Robson lives with his family in Wilmington, Delaware.